Selected Titles in This Series

D1786598

711 **V. G. Kac, C. Martinez, and E. Zelmanov,** Graded simple Jordan superalgebras of growth one, 2001

710 **Brian Marcus and Selim Tuncel,** Resolving Markov chains onto Bernoulli shifts via positive polynomials, 2001

709 **B. V. Rajarama Bhat,** Cocylces of CCR flows, 2001

708 **William M. Kantor and Ákos Seress,** Black box classical groups, 2001

707 **Henning Krause,** The spectrum of a module category, 2001

706 **Jonathan Brundan, Richard Dipper, and Alexander Kleshchev,** Quantum Linear groups and representations of $GL_n(\mathbb{F}_q)$, 2001

705 **I. Moerdijk and J. J. C. Vermeulen,** Proper maps of toposes, 2000

704 **Jeff Hooper, Victor Snaith, and Min van Tran,** The second Chinburg conjecture for quaternion fields, 2000

703 **Erik Guentner, Nigel Higson, and Jody Trout,** Equivariant E-theory for C^*-algebras, 2000

702 **Ilijas Farah,** Analytic guotients: Theory of liftings for quotients over analytic ideals on the integers, 2000

701 **Paul Selick and Jie Wu,** On natural coalgebra decompositions of tensor algebras and loop suspensions, 2000

700 **Vicente Cortés,** A new construction of homogeneous quaternionic manifolds and related geometric structures, 2000

699 **Alexander Fel′shtyn,** Dynamical zeta functions, Nielsen theory and Reidemeister torsion, 2000

698 **Andrew R. Kustin,** Complexes associated to two vectors and a rectangular matrix, 2000

697 **Deguang Han and David R. Larson,** Frames, bases and group representations, 2000

696 **Donald J. Estep, Mats G. Larson, and Roy D. Williams,** Estimating the error of numerical solutions of systems of reaction-diffusion equations, 2000

695 **Vitaly Bergelson and Randall McCutcheon,** An ergodic IP polynomial Szemerédi theorem, 2000

694 **Alberto Bressan, Graziano Crasta, and Benedetto Piccoli,** Well-posedness of the Cauchy problem for $n \times n$ systems of conservation laws, 2000

693 **Doug Pickrell,** Invariant measures for unitary groups associated to Kac-Moody Lie algebras, 2000

692 **Mara D. Neusel,** Inverse invariant theory and Steenrod operations, 2000

691 **Bruce Hughes and Stratos Prassidis,** Control and relaxation over the circle, 2000

690 **Robert Rumely, Chi Fong Lau, and Robert Varley,** Existence of the sectional capacity, 2000

689 **M. A. Dickmann and F. Miraglia,** Special groups: Boolean-theoretic methods in the theory of quadratic forms, 2000

688 **Piotr Hajłasz and Pekka Koskela,** Sobolev met Poincaré, 2000

687 **Guy David and Stephen Semmes,** Uniform rectifiability and quasiminimizing sets of arbitrary codimension, 2000

686 **L. Gaunce Lewis, Jr.,** Splitting theorems for certain equivariant spectra, 2000

685 **Jean-Luc Joly, Guy Metivier, and Jeffrey Rauch,** Caustics for dissipative semilinear oscillations, 2000

684 **Harvey I. Blau, Bangteng Xu, Z. Arad, E. Fisman, V. Miloslavsky, and M. Muzychuk,** Homogeneous integral table algebras of degree three: A trilogy, 2000

683 **Serge Bouc,** Non-additive exact functors and tensor induction for Mackey functors, 2000

682 **Martin Majewski,** ational homotopical models and uniqueness, 2000

(Continued in the back of this publication)

Graded Simple Jordan Superalgebras of Growth One

MEMOIRS
of the
American Mathematical Society

Number 711

Graded Simple Jordan Superalgebras of Growth One

V. G. Kac
C. Martinez
E. Zelmanov

March 2001 • Volume 150 • Number 711 (second of 5 numbers) • ISSN 0065-9266

American Mathematical Society
Providence, Rhode Island

2000 *Mathematics Subject Classification.*
Primary 17C70, 17B70; Secondary 17B60, 17B65, 17B66, 17B68.

Library of Congress Cataloging-in-Publication Data

Kac, Victor G., 1943–
Graded simple Jordan superalgebras of growth one / V. G. Kac, C. Martinez, E. Zelmanov.
p. cm. — (Memoirs of the American Mathematical Society, ISSN 0065-9266 ; no. 711)
"Volume 150, number 711 (second of 5 numbers).
Includes bibliographical references.
ISBN 0-8218-2645-X
1. Jordan algebras. 2. Superalgebras. I. Martinez, C. (Consuelo), 1955– II. Zelmanov, Efim, 1955– III. Title. IV. Series.

QA3.A57 no. 711
[QA252.5]
510 s—dc21
[512′.24] 00-053582

Memoirs of the American Mathematical Society

This journal is devoted entirely to research in pure and applied mathematics.

Subscription information. The 2001 subscription begins with volume 149 and consists of six mailings, each containing one or more numbers. Subscription prices for 2001 are \$494 list, \$395 institutional member. A late charge of 10% of the subscription price will be imposed on orders received from nonmembers after January 1 of the subscription year. Subscribers outside the United States and India must pay a postage surcharge of \$31; subscribers in India must pay a postage surcharge of \$43. Expedited delivery to destinations in North America \$35; elsewhere \$130. Each number may be ordered separately; *please specify number* when ordering an individual number. For prices and titles of recently released numbers, see the New Publications sections of the *Notices of the American Mathematical Society.*

Back number information. For back issues see the *AMS Catalog of Publications.*

Subscriptions and orders should be addressed to the American Mathematical Society, P. O. Box 845904, Boston, MA 02284-5904. *All orders must be accompanied by payment.* Other correspondence should be addressed to Box 6248, Providence, RI 02940-6248.

Memoirs of the American Mathematical Society is published bimonthly (each volume consisting usually of more than one number) by the American Mathematical Society at 201 Charles Street, Providence, RI 02904-2294. Periodicals postage paid at Providence, RI. Postmaster: Send address changes to Memoirs, American Mathematical Society, P. O. Box 6248, Providence, RI 02940-6248.

This publication is indexed in *Science Citation Index*®, *SciSearch*®, *Research Alert*®, *CompuMath Citation Index*®, *Current Contents*®/*Physical, Chemical & Earth Sciences.*
Printed in the United States of America.

∞ The paper used in this book is acid-free and falls within the guidelines established to ensure permanence and durability.
Visit the AMS home page at URL: `http://www.ams.org/`

10 9 8 7 6 5 4 3 2 1 06 05 04 03 02 01

Contents

Abstract

We classify graded simple Jordan superalgebras of growth one which correspond to the so called "superconformal algebras" via the Tits-Kantor-Koecher construction.

The superconformal algebras with a "hidden" Jordan structure are those of type K and the recently discovered Cheng-Kac superalgebras $CK(6)$. We show that Jordan superalgebras related to the type K are Kantor Doubles of some Jordan brackets on associative commutative superalgebras and list these brackets.

Received by the editor May 26, 1998.

1991 *Mathematics Subject Classification.* Primary 17C70, 17B70; Secondary 17B60, 17B65, 17B66, 17B68.

Key words and phrases. superalgebra, Jordan algebra, superconformal algebra, bracket, Kantor double.

The first author was partially supported by NSF grant DMS-9622870.

The second author was partially supported by FICYT PGI-PB99-04, FEDER IFD-97-0556 and DGES PB97-1291-C03-01 .

The third author was partially supported NSF grant DMS-9704132.

Introduction

Statement of the Problem

Superconformal algebras (such as the superalgebras of Ramond, Neveu-Schwarz, etc.; see [Ra], [NS]) are $\mathbb{Z}$-graded simple Lie superalgebras, $L = \sum_{i \in \mathbb{Z}} L_i$ over an algebraically closed field F of zero characteristic, containing the Virasoro algebra in the even part and such that the dimensions $dim(L_i)$, $i \in \mathbb{Z}$, are uniformely bounded.

In [KL] V. Kac and J. van de Leur conjectured that Lie superalgebras of this type admit a classification that is similar to the classification of $\mathbb{Z}$-graded simple Lie algebras obtained by V. Kac [K1] and O. Mathieu [Ma1, Ma2]. To be more precise, let $F[t^{-1}, t, \xi_1, \ldots, \xi_n]$ be the associative commutative superalgebra of polynomials in one Laurent variable t and n odd variables $\xi_1, \ldots, \xi_n$. The Lie superalgebra $W(1,n)$ of derivations of the superalgebra $F[t^{-1}, t, \xi_1, \ldots, \xi_n]$, $\mathbb{Z}$-graded by degrees of t, is a $\mathbb{Z}$-graded simple Lie superalgebra containing the Virasoro algebra Vir in the even part and having dimensions of all homogeneous components uniformely bounded. Kac and van de Leur conjectured that an arbitrary $\mathbb{Z}$-graded simple Lie superalgebra containing Vir in the even part and having dimensions of all homogeneous components uniformely bounded is isomorphic to $W(1,n)$ (for some n) or to one of known subalgebras of $W(1,n)$ of types S and K. However, one more example of a superconformal algebra was discovered, the exceptional Lie superalgebra CK(6) (see [CK]). Furthermore, it has been establish recently, using methods of [AK] (where the non-super case is treated), that the Locality Principle and a natural finiteness condition suffice to show that the above list is complete ([K5],[K6]).

Dropping the assumption of the existence of a Virasoro subalgebra brings in two more classes: (twisted) loop algebras and superalgebras of Cartan type, and it is natural to expect that there will be no new surprises.

The present paper is the first step towards the proof of this conjecture. The basic idea is the following. The zero component L_0 of such a $\mathbb{Z}$-graded Lie superalgebra is a finite dimensional Lie superalgebra, $L_0 = L_{0\bar{0}} + L_{0\bar{1}}$. It is known (see [K2]) that a finite dimensional Lie superalgebra is solvable if and only if its even part is solvable. Suppose that the Lie algebra $L_{0\bar{0}}$ is not solvable. Then (see [J2]) $L_{0\bar{0}}$ contains a subalgebra $sl_2(F) = Fe + Fh + Ff$ with $[e,f] = h$, $[f,h] = 2f$, $[e,h] = -2e$.

An arbitrary homogeneous component L_i is a module over $sl_2(F)$. Since there is only one irreducible $sl_2(F)$-module in each dimension and dimensions of L_i are uniformely bounded it follows that only finitely many irreducible $sl_2(F)$-modules can occur in decompositions of $L_i,\ i \in \mathbb{Z}$. This implies that $ad(h) : L \longrightarrow L,\ x \longrightarrow [x, h]$ is diagonalizable and has finitely many eigenvalues. It was shown in [Z5] that such Lie algebras can be studied by means of Jordan theory.

Definitions and Notation

In the present paper we give a complete classification in the simplest case when $ad(h)$ has only two non-zero eigenvalues. In this case the classification of Lie superalgebras in question is implied by the classification of $\mathbb{Z}$-graded simple Jordan superalgebras $J = \sum_{i\in\mathbb{Z}} J_i$ with $\dim J_i$ uniformely bounded.

A (linear) Jordan algebra is a vector space J with a binary operation $(x, y) \longrightarrow xy$ satisfying the following identities:
(J1) $xy = yx$
(J2) $(x^2 y)x = x^2(yx)$.
For an element $x \in J$ let $R(x)$ denote the right multiplication $R(x) : a \longrightarrow ax$ in J. Then linearization of (J2) yields the following linearization in the operator form:

$$R((xy)z) + R(x)R(z)R(y) + R(y)R(z)R(x) =$$

$$R(xy)R(z) + R(xz)R(y) + R(yz)R(x) =$$

$$R(z)R(xy) + R(y)R(xz) + R(x)R(yz).$$

We will refer to it as "the Jordan identity".

For elements x, y, z of a Jordan algebra J by $\{x, y, z\}$ we denote their Jordan triple product, $\{x, y, z\} = (xy)z + x(yz) - y(xz)$.
By $U(x, y),\ V(x, y)$ we denote the operators $U(x, y) : J \longrightarrow J$, $U(x, y) : z \longrightarrow \{x, z, y\}$ and $V(x, y) : J \longrightarrow J,\ V(x, y) : z \longrightarrow \{z, x, y\}$ respectively. Let $U(x) = U(x, x)$.
For arbitrary $x, y \in J$ the operator $D(x, y) = R(x)R(y) - R(y)R(x)$ is known to be a derivation of J (see [J3]).
We will need also the following identity:

$$R(x)R(y)R(z) = \tfrac{1}{2}(-R((xz)y) + R(xy)R(z) + R(xz)R(y) + R(yz)R(x) - R(x)D(z, y) - R(z)D(y, x) - R(yD(z, x)) + R(y)D(z, x). \qquad \textbf{(D)}$$

Let $End_F J$ be the associative algebra of all linear transformations on the vector space J. Let $R < J >$ be the subalgebra of $End_F J$ generated by all multiplication operators $R(x),\ x \in J$. We will refer to $R < J >$ as the multiplication algebra of J. If $1 \in J$, then from the identity **(D)** it follows that $R < J >$ is the linear span of operators of the type $R(a)R(b)D_1 \cdots D_r$, where $a, b \in J;\ D_1, \ldots, D_r \in D(J, J)$ are inner derivations.

Let's discuss briefly a relation between Lie algebras and Jordan algebras (in fact, one of many). Let L be a Lie algebra containing a subalgebra $Fe + Fh + Ff$ which is isomorphic to $sl_2(F)$, that is, $[e, f] = h$, $[f, h] = 2f$, $[e, h] = -2e$.

Suppose that the operator $ad(h) : L \to L$ is diagonalizable and that the only eigenvalues of $ad(h)$ are -2,0,2. Let $L = L_{(-2)} + L_{(0)} + L_{(2)}$ be the decomposition of L into the sum of eigenspaces. Following J. Tits [T] we define a structure of a Jordan algebra on $L_{(-2)}$ via $x_{(-2)} \star y_{(-2)} = [[x_{(-2)}, f], y_{(-2)}]$ for arbitrary elements $x_{(-2)}, y_{(-2)} \in L_{(-2)}$. On the other hand, for an arbitrary Jordan algebra J with 1 there exists a unique (up to isomorphism) pair $L \supseteq sl_2(F)$ with these properties such that $L_{(-2)} \simeq J$ and L has zero center. We will call such a Lie algebra the Tits-Kantor-Koecher construction of J and denote it $K(J) = J^- + [J^-, J^+] + J^+$ (see [Ka1,Ko,T]).

Let $\mathcal{V}$ be a homogeneous variety of algebras, that is, a class of F-algebras satisfying a certain set of homogeneous identities (see [ZSSS]). Let $A \in \mathcal{V}$.

Let M be a vector space with two bilinear mappings $f : M \times A \longrightarrow M$ and $g : A \times M \longrightarrow M$.

Consider the direct sum of vector spaces $A + M$. For arbitrary elements $a, b \in A$; $m, m' \in M$ let $a \cdot b = ab$, $a \cdot m = g(a, m)$, $m \cdot a = f(m, a)$, $m \cdot m' = 0$. Following S. Eilenberg (see [J3]) we call M a $\mathcal{V}$-bimodule over A if $(A + M, \cdot) \in \mathcal{V}$. Thus we can speak about Lie bimodules, Jordan bimodules, etc.

Let us introduce the definition of a superalgebra corresponding to a variety $\mathcal{V}$. In general, by a superalgebra we mean just a $\mathbb{Z}/2\mathbb{Z}$-graded algebra, $A = A_{\bar{0}} + A_{\bar{1}}$.

Example Let V be a vector space. The Grassmann (or exterior) algebra $G(V)$ is the quotient of the tensor algebra $T(V)$ modulo the ideal generated by symmetric tensors $v \otimes w + w \otimes v$; $v, w \in V$. Clearly $G(V) = G_{\bar{0}} + G_{\bar{1}}$, where $G_{\bar{0}}$ (resp. $G_{\bar{1}}$) is spanned by products of elements of V of even (resp. odd) length.

Let V be a vector space of countable dimension. By the Grassmann envelope of a superalgebra $A = A_{\bar{0}} + A_{\bar{1}}$ we mean the subalgebra $G(A) = A_{\bar{0}} \otimes G_{\bar{0}} + A_{\bar{1}} \otimes G_{\bar{1}}$ of the tensor product $A \otimes G(V)$.

Definition A superalgebra $A = A_{\bar{0}} + A_{\bar{1}}$ is called a $\mathcal{V}$-superalgebra if the Grassmann envelope $G(A)$ lies in $\mathcal{V}$.

In particular, if $A = A_{\bar{0}} + A_{\bar{1}}$ is a $\mathcal{V}$- superalgebra, then $A_{\bar{0}} \in \mathcal{V}$ and $A_{\bar{1}}$ is a $\mathcal{V}$-bimodule over $A_{\bar{0}}$.

In this way one can define Lie superalgebras, Jordan superalgebras, etc. Clearly, associative superalgebras are just $\mathbb{Z}/2\mathbb{Z}$-graded associative algebras and a commutative superalgebra is more often called a supercommutative (super)algebra.

Everything that we said about Jordan algebras can be extended to Jordan superalgebras in a straightforward way.

The Main Result

Now let us return to a graded simple Lie superalgebra, $L = \sum_{i \in \mathbb{Z}} L_i$ such that $dim(L_i)$ are uniformely bounded and $L_{0\bar{0}}$ contains a subalgebra $Fe + Fh + Ff \simeq$

$sl_2(F)$. Suppose, as above, that the set of eigenvalues of $ad(h)$ is minimal, that is, the eigenvalues of $ad(h)$ are -2, 0, 2. Let $L = L_{(-2)} + L_{(0)} + L_{(2)}$ be a decomposition of L in the sum of eigenspaces, so that $f \in L_{(2)}$, $e \in L_{(-2)}$. As above, for arbitrary elements $x_{(-2)}, y_{(-2)} \in L_{(-2)}$ we let $x_{(-2)} \star y_{(-2)} = [[x_{(-2)}, f], y_{(-2)}] \in L_{(-2)}$. Then $J = (L_{(-2)}, \star)$ becomes a Jordan superalgebra and $L \simeq K(J)$.

The Jordan superalgebra $J = L_{(-2)}$ is $\mathbb{Z}$-graded simple, $J = \sum_{i\in\mathbb{Z}} J_i$ and $dim(J_i) = dim(L_{(-2)} \cap L_i) < dim(L_i)$, so the dimensions $dim J_i$ are uniformely bounded.

The aim of this paper is to classify graded simple Jordan superalgebras with this property. Let us start with some examples.

1) Let n be a natural number, $\mathcal{G}$ be a $\mathbb{Z}/n\mathbb{Z}$-graded finite dimensional superalgebra, $\mathcal{G} = \mathcal{G}_0 + \mathcal{G}_1 + \cdots + \mathcal{G}_{n-1}$. For an arbitrary integer i, let $\bar{i}$, $0 \leq \bar{i} \leq n-1$, denote the residue of i modulo n. By a (twisted) loop algebra corresponding to $\mathcal{G}$ we mean the subalgebra $\mathcal{L}(G) = \sum_{i\in\mathbb{Z}} \mathcal{G}_{\bar{i}} \otimes t^i$ of the tensor product $\mathcal{G} \otimes F[t^{-1}, t]$. If $\mathcal{G}$ is a finite dimensional simple Jordan (Lie) superalgebra, $\mathcal{L}(G)$ is a graded simple Jordan (Lie) superalgebra having all dimensions $dim(\mathcal{L}(G)_i$, $i \in \mathbb{Z}$, uniformely bounded.

2) Let V be a direct sum of vector spaces, both $V_{\bar{0}} = \oplus_{i\in\mathbb{Z}} V_{\bar{0}i}$ and $V_{\bar{1}} = \oplus_{i\in\mathbb{Z}} V_{\bar{1}i}$ are represented as direct sums of finite dimensional vectors spaces such that dimensions of subspaces $V_i = V_{\bar{0}i} + V_{\bar{1}i}$, $i \in \mathbb{Z}$, are uniformely bounded. Suppose further that the space V is equipped with a nondegenerate supersymmetric bilinear form $<,>: V \times V \to F$. Supersymmetric here means that the form $<,>$ is symmetric on $V_{\bar{0}}$, skew-symmetric on $V_{\bar{1}}$ and $< V_{\bar{0}}, V_{\bar{1}} > = < V_{\bar{1}}, V_{\bar{0}} > = (0)$. Then the direct sum of vector spaces $J = F1 + V = J_{\bar{0}} + J_{\bar{1}}$, $J_{\bar{0}} = F1 + V_{\bar{0}}$, $J_{\bar{1}} = V_{\bar{1}}$ becomes a Jordan superalgebra under multiplication $vw = < v, w > 1$ for $v, w \in V$. The superalgebra J is simple and graded, $J_i = V_i$ for $i \neq 0$, $J_0 = F1 + V_0$.

We shall often say superform in place of a nondegenerate supersymmetric bilinear form.

3) Let $A = A_{\bar{0}} + A_{\bar{1}}$ be an associative commutative superalgebra. If $a \in A_{\bar{i}}$ then we denote $|a| = i$. By a bracket on A we mean a bilinear mapping $\{,\} : A \times A \to A$.

Starting with a bracket $\{,\}$ on A consider a direct sum of vector spaces $J = J(A, \{,\}) = A + Ax$. We shall define a multiplication on J. For arbitrary elements $a, b \in A$ their product in J is the product ab in A and $a(bx) = (ab)x$, $(bx)a = (-1)^{|a|}(ba)x$, $(ax)(bx) = (-1)^{|b|}\{a, b\}$.

The $\mathbb{Z}/2\mathbb{Z}$-gradation on A can be extended to a $\mathbb{Z}/2\mathbb{Z}$-gradation on J via $J_{\bar{0}} = A_{\bar{0}} + A_{\bar{1}}x$, $J_{\bar{1}} = A_{\bar{1}} + A_{\bar{0}}x$. We call J the Kantor Double of $(A, \{,\})$.

A bracket $\{,\}$ on A is called a Jordan bracket if the Kantor Double $J(A, \{,\})$ is a Jordan superalgebra.

Example. A bracket $\{,\} : A \times A \to A$ is called a Poisson bracket if $(A, \{,\})$ is a Lie superalgebra and $\{ab, c\} = a\{b, c\} + (-1)^{|b||c|}\{a, c\}b$ for arbitrary elements $a, b, c \in A$. An arbitrary Poisson bracket is a Jordan bracket (see [Ka2]).

It immediatly follows from the Jordan identity that for an arbitrary Jordan bracket $\{,\}$

(1) $D : a \to \{a, 1\}$ is a derivation of A,

(2) $\{a, bc\} = \{a, b\}c + (-1)^{|a||b|}b\{a, c\} - D(a)bc.$

(For other properties of Jordan brackets see [KM]). If the superalgebra A is generated by elements $\{a_i\}_i$ then a Jordan bracket $\{,\}$ is determined by the derivation D and by values $\{a_i, a_j\}$.

Examples. Let $A = F[t^{-1}, t, \xi_1, \ldots, \xi_n]$.

1) $D = \frac{\partial}{\partial t}t$, $[\xi_i, \xi_j] = -\delta_{ij}$, $[t, \xi_i] = 0$.

2) $D = \frac{\partial}{\partial t}$, $[\xi_i, \xi_j] = -\delta_{ij}$, $[t, \xi_i] = 0$.

We will refer to the brackets 1), 2) as the Jordan brackets of Neveu-Schwarz and Ramond types respectively (see [NS], [R]).

Let $n \geq 1$, $V = V_0 + \ldots + V_{n-1}$ be a finite dimensional $\mathbb{Z}/n\mathbb{Z}$-graded vector space over F. The gradation on V can be uniquely extended to a $\mathbb{Z}/n\mathbb{Z}$-gradation on the Grassmann algebra, $G(V) = \sum_{i=0}^{n-1} G(V)_i$.

Let $A = \sum_{i \in \mathbb{Z}} G(V)_{\bar{i}} \otimes t^i = \mathcal{L}(G(V))$. If $[,] : A \times A \to A$ is a Jordan bracket, the element x in the Kantor Double construction is given degree i such that $2i \in n\mathbb{Z}$ and $[A_j, A_k] \subseteq A_{j+k+2i}$ for arbitrary $j, k \in \mathbb{Z}$, then the Kantor Double $J = A + Ax$ is a $\mathbb{Z}$-graded Jordan superalgebra having all dimensions $dim(J_i)$, $i \in \mathbb{Z}$, uniformely bounded

4) We say that a graded simple Jordan superalgebra J is of Cartan type if J contains a graded subalgebra B of finite codimension such that the corresponding subspace $B^- + [B^-, J^+] + [J^-, B^+] + B^+$ of the Tits-Kantor-Koecher Lie superalgebra $K(J)$ is a subalgebra of $K(J)$ of finite codimension. Thus $K(J)$ is a Lie superalgebra of Cartan type (see [K2]).

Clearly for Jordan superalgebras J of examples 1),2), 4) their Lie superalgebras $K(J)$ do not contain the Virasoro algebra.

The main result of this paper is the following theorem:

Theorem Let $J = \sum_{i \in \mathbb{Z}} J_i$ be an infinite dimensional $\mathbb{Z}$-graded simple Jordan superalgebra with 1 over an algebraically closed field F of zero characteristic such that dimensions of J_i are uniformely bounded. Then J is isomorphic to one of the following superalgebras:

1) a loop superalgebra $\mathcal{L}(G)$, where $\mathcal{G} = \mathcal{G}_0 + \mathcal{G}_1 + \cdots + \mathcal{G}_{n-1}$ is a finite dimensional simple $\mathbb{Z}/n\mathbb{Z}$- graded Jordan superalgebra,

2) a Jordan superalgebra $F1 + V$ of a nondegenerate supersymmetric form in a $\mathbb{Z}$-graded vector space $V = V_{\bar{0}} + V_{\bar{1}}$,

3) a Kantor Double $J = A + Ax$ of an associative commutative superalgebra $A = \sum_{i \in \mathbb{Z}} G(V)_{\bar{i}} \otimes t^i$ with a Jordan bracket, where $V = V_0 + \cdots + V_{n-1}$ is a $\mathbb{Z}/n\mathbb{Z}$- graded finite dimensional vector space. If n is odd then there is one (up to isomorphism) Jordan superalgebra of this type with x being of degree 0. If n is even then there are two Jordan brackets on A, of Neveu-Schwarz and Ramond types respectfully (see Examples 1, 2 and Chapter 6) leading to two nonisomorphic Jordan superalgebras, one with x having degree 0 and one with x having degree $\frac{-n}{2}$,

4) a Jordan superalgebra of Cartan type,

5) two exceptional Jordan superalgebra whose Tits-Kantor Koecher constructions are isomorphic to the exceptional superalgebras $CK(6)$ of Neveu-Schwarz and Ramond types (see [CK]).

It is easy to see that if J is a (twisted) loop Jordan superalgebra then its Tits-Kantor-Koecher Lie superalgebra $K(J)$ is a (twisted) loop Lie superalgebra. If $J = F1 + V$ is a Jordan superalgebra of a superform and the space V is infinite dimensional then the zero component of $K(J)$ is infinite dimensional. If J is a Kantor Double corresponding to a Jordan bracket of Neveu-Schwarz (respectively Ramond) type, then $K(J)$ is a superconformal algebra of K of Neveu-Schwarz (respectively Ramond) type. If J is a superalgebra of Cartan type then so is $K(J)$. In this case $K(J)$ is isomorphic to the positive part of one of the algebras K_n, $CK(6)$ of Neveu-Schwarz type (see [K7]).

Finally, the exceptional Jordan superalgebras correspond to the superconformal algebras $CK(6)$.

Thus the theorem confirms the Kac-van de Leur conjecture as modified above.

Structure of the Proof

Let $J = J_{\bar{0}} + J_{\bar{1}}$ be a Jordan superalgebra satisfying the assumptions of the theorem. For the sake of convenience we will denote the Jordan algebra $J_{\bar{0}}$ as A and the Jordan bimodule $J_{\bar{1}}$ over A as M, $J = A + M$.

In Chapter 1 we determine possible structures of the Jordan algebra A.

Along with

(a) finite dimensional simple Jordan algebras;

(b) loop algebras $\mathcal{L}(G)$ of finite dimensional simple $\mathbb{Z}/n\mathbb{Z}$- graded Jordan algebras $\mathcal{G} = \mathcal{G}_0 + \mathcal{G}_1 + \cdots + \mathcal{G}_{n-1}$;

(c) infinite dimensional Jordan algebras of a symmetric nondegenerate bilinear form $J = F1 + V$, $V = \oplus_{i\in\mathbb{Z}} V_i$,
consider

(d) a graded subalgebra $A \subseteq \mathcal{L}(G)$ such that only finitely many negative components A_i, $i < 0$, are nonzero and there exists $k \geq 1$ such that $A \supseteq \sum_{i=k}^{\infty} \mathcal{L}(G)_i$. In this case we say that A is **positively graded**. Similarly one can define a **negatively graded** algebra as a graded subalgebra $A \subseteq \mathcal{L}(G)$ having only finitely many nonzero positive components and such that $A \supseteq \sum_{i<k} \mathcal{L}(G)_i$, for some $k < 0$.

Recall that an arbitrary algebra R is said to be **prime** if for arbitrary nonzero ideals I_1, I_2 of R we have $I_1 I_2 \neq (0)$.

A nonzero element a of a Jordan algebra A is called an absolute zero divisor if $a^2 = 0$ and $\{a, A, a\} = (0)$. A Jordan algebra is said to be **nondegenerate** if it does not contain absolute zero divisors. The smallest ideal $M(A)$ of A whose quotient algebra $A/M(A)$ is nondegenerate is called the McCrimmon radical of A.

In [MZ2] it was proved that a prime nondegenerate $\mathbb{Z}$-graded Jordan algebra $A = \oplus_{i\in\mathbb{Z}} A_i$ such that the dimensions $dim A_i$, $i \in \mathbb{Z}$ are uniformely bounded is of one of the types (a), (b), (c) or (d).

For an arbitrary odd element $x \in M$ the even operator $R(x)^2 : J \to J$ is a derivation of J. Let $\mathcal{D}$ denote the F-linear span of all derivations $R(x)^2$, $x \in M$. Clearly, $\mathcal{D}$ is closed under commutation and thus is a Lie algebra.

The algebra A contains the largest $\mathcal{D}$-invariant ideal I such that the algebra I_0 is nilpotent. In Chapter 1 we prove the following result.

Proposition If a superalgebra J satisfies the assumptions of the Theorem then one of the following assertions holds:

1) $A/I \simeq \mathcal{L}(G)$ a loop algebra of a simple finite dimensional Jordan algebra of a bilinear form, the ideal I is nilpotent and $I \neq (0)$,

2) A/I is a one-sided graded algebra commensurable with a loop algebra $\mathcal{L}(G)$ of a simple finite dimensional Jordan algebra of a bilinear form,

3) A/I is a finite dimensional simple Jordan algebra of a bilinear form, $I \neq (0)$,

4) $A = A^{(1)} \oplus A^{(2)}$, where $A^{(i)}$ are algebras of the types (a), (b), (c) or (d) that we have mentioned above,

5) $A \simeq \mathcal{L}(G)$, where $\mathcal{G}$ is a simple finite dimensional Jordan algebra,

6) A is a finite dimensional simple algebra,

7) A/I is a simple infinite dimensional Jordan algebra of a bilinear form, $I \neq (0)$,

8) A is a simple infinite dimensional Jordan algebra of a bilinear form.

In Chapter 2 we analize the cases when A is one-sided graded (case 2), or A is a direct sum of two positively (negatively) graded algebras (part of case 4) or A is a direct sum of a finite dimensional algebra and a one-sided graded algebra (part of case 4) and prove that in all those cases the superalgebra J is of Cartan type.

In Chapter 3 we assume A to be the algebra of Laurent polynomials (part of case 5) or a direct sum of two loop algebras (part of case 4). In these cases J is a loop superalgebra or a Kantor Double.

In Chapter 4 we finish the case when A is a loop algebra of a simple finite dimensional Jordan algebra (case 5). We prove that J is either a loop superalgebra or the exceptional Jordan superalgebra J_8.

In Chapter 5 we consider the case when A is isomorphic to an infinite dimensional Jordan algebra $A = F1 + V$ of a symmetric nondegenerate bilinear form, $V = \sum_{i \in \mathbb{Z}} V_i$ (case 8) or A/I is finite dimensional ($I = (0)$ or $I \neq 0$) (cases 3 and 6) and prove that J is either a Jordan algebra of a superform or is finite dimensional.

In Chapter 6 we consider the (main) case when $I \neq (0)$ and A/I is isomorphic to a loop algebra of a simple finite dimensional Jordan algebra (case 1). In this case J is a Kantor double.

Finally in Chapter 7 we consider the remaining cases and show that they are impossible.

Apart from connections to superconformal algebras there are inner reasons in Jordan theory for an interest in superalgebras $J(A, [,])$ of Jordan brackets. The key tool in classification of prime nondegenerate Jordan algebras (see [Z3]) was the notion of a "tetrad eater", a Jordan polynomial $f(x_1, \ldots, x_n)$ such that $fx_{n+1}x_{n+2}x_{n+3} + x_{n+3}x_{n+2}x_{n+1}f$ is again a Jordan polynomial.

It seems natural to conjecture that a Jordan polynomial is a tetrad eater" if and only if it is identically zero in Grassmann envelopes of all Kantor Doubles $J(A; \{,\})$, where A is an associative commutative superalgebra with a Jordan bracket.

CHAPTER 1

Structure of the Even Part

1.1. General results

Let $J = A + M = \sum_{i \in Z} J_i$ be a graded simple Jordan superalgebra with 1 such that dimensions dim J_i, $i \in \mathbb{Z}$, are uniformely bounded. Let $\mathcal{D}$ be the F-linear span of all derivations $R(x)^2$, $x \in M$.

If I', I'' are $\mathcal{D}$-invariant graded ideals of A such that subalgebras I'_0, I''_0 are nilpotent, then the ideal $I' + I''$ is $\mathcal{D}$-invariant and the subalgebra $(I' + I'')_0 = I'_0 + I''_0$ is nilpotent. This implies that there exists the largest $\mathcal{D}$-invariant ideal I of A such that I_0 is nilpotent.

Let R be an associative algebra. If we replace the multiplication on R by the symmetric multiplication $a \circ b = \frac{1}{2}(ab + ba)$ then we get a structure of a Jordan algebra on R which will be denoted as $R^{(+)}$.

For a subset X of R let $< X >$ denote the subalgebra of R generated by X.

An algebra is said to be locally nilpotent if every finitely generated subalgebra of it is nilpotent.

LEMMA 1.1. *Let $R = \sum_{i \in \mathbb{Z}} R_i$ be a graded associative algebra, J a finitely generated graded Jordan subalgebra of $R^{(+)}$ such that dimensions dimJ_i, $i \in \mathbb{Z}$ are uniformely bounded. Suppose further that $R =< J >$ and that the algebra J_0 is nilpotent. Then the algebra R_0 is locally nilpotent.*

PROOF. Let $Loc(R)$ denote the locally nilpotent radical of the algebra R, that is, the largest ideal of R that is locally nilpotent (see [J1,R1]). Since $Loc(R)$ is clearly invariant with respect to all automorphisms of the algebra R it follows that $Loc(R)$ is a graded ideal. Suppose that the algebra R_0 is not locally nilpotent. Then, without loss of generality, we can assume that the algebra R is prime and $Loc(R) = (0)$. Indeed, let S be a finitely generated graded subalgebra of R_0. Suppose that S is not nilpotent. Let P be a maximal graded ideal of R such that S is not nilpotent modulo P. Then (see [ZSSS]) the quotient algebra R/P is graded prime and $Loc(R/P) = (0)$. Now it remains to consider R/P and $J + P/P$ instead of R and J.

Since the algebra R is prime and $Loc(R) = (0)$ it follows (see [Z2]) that the algebra J is prime and nondegenerate. By the classification theorem in [MZ2] the algebra J is either (i) finite dimensional, or (ii) a loop algebra, or (iii) an infinite dimensional Jordan algebra of a nondegenerate symmetric bilinear form or (iv) a positively graded or a negatively graded algebra.

The cases (i), (ii) and (iii) are impossible because in all these cases J_0 contains an identity. Suppose therefore that the algebra J is positively graded (the case of J being negatively graded is similar).

Let $J = J_{-k} + J_{-(k-1)} + \ldots + J_0 + J_1 + \ldots$ The subalgebra $J_{-k} + \ldots + J_{-1}$ of J is nilpotent of degree $\leq k+1$. Hence the Jordan algebra $J_{-k} + \ldots + J_{-1} + J_0$ is also nilpotent and the associative subalgebra $R_{(-)}$ of R generated by $J_{-k} + \ldots + J_{-1} + J_0$ is nilpotent and finite dimensional (see [ZSSS]).

Let $R_{(+)}$ denote the subalgebra of R generated by $\sum_{i\geq 1} J_i$. We have $R = R_{(-)} + R_{(-)}R_{(+)} + R_{(+)}$. This implies that there exists $l \geq 1$ such that $R_i = (0)$ for $i < -l$. Hence the ideal $id_R(\sum_{i\geq 2l+1} R_i)$ of the algebra R generated (as an ideal) by $\sum_{i\geq 2l+1} R_i$ lies in $R_{(+)}$.

We claim that the algebra $\bar{R} = R/id_R(\sum_{i\geq 2l+1} R_i)$ is nilpotent.

Indeed, the algebra $\bar{R}$ is generated by the image $\bar{J}$ of the algebra J. We have $\bar{J} = \bar{J}_{-k} + \ldots + \bar{J}_{2l}$. Hence, the algebra $\bar{J}$ is finite dimensional. An arbitrary homogeneous element of $\bar{J}$ is nilpotent. By the Wedderburn theorem for Jordan algebras (see [Sh]) the algebra $\bar{J}$ is nilpotent. Since the algebra $\bar{R}$ is generated by $\bar{J}$ it follows (see [ZSSS]) that $\bar{R}$ is also nilpotent.

Let $\bar{R}^s = (0)$. Then $R_0^s \subseteq id(\sum_{i\geq 2l+1} R_i) \subseteq R_{(+)}$, which implies that $R_0^s = (0)$, the contradiction. Lemma is proved. □

Denote $\bar{A} = A/I$. An arbitrary nonzero $\mathcal{D}$-invariant graded ideal of $\bar{A}$ has a nonnilpotent intersection with $\bar{A}_0$ (because of maximality of I).

DEFINITION 1.2. We call a nonzero $\mathcal{D}$-invariant ideal P of $\bar{A}$ 0-minimal if for every nonzero $\mathcal{D}$-invariant graded ideal P' of $\bar{A}$ which is contained in P we have $P'_0 = P_0$.

LEMMA 1.3. *Let P be a 0-minimal ideal of $\bar{A}$. Then the algebra P is nondegenerate and prime.*

PROOF. The McCrimmon radical $\mathrm{M}(\bar{A})$ is $\mathcal{D}$-invariant, graded and its zero component is nilpotent. Hence the algebra $\bar{A}$ is nondegenerate. This implies that P is nondegenerate.

Let K and L be nonzero graded ideals of P such that $K \cdot L = (0)$. Consider the descending chain $K^{<0>} = K$, $K^{<i+1>} = (K^{<i>})^3$.

For an arbitrary subset $B \subseteq \bar{A}$ its annihilator $Ann_{\bar{A}}B$ in $\bar{A}$ is definitionined as follows

$$Ann_{\bar{A}}B = \{a \in \bar{A} | B \cdot a = (0),\ \bar{A}D(B,a) = (0)\ \}.$$

For an arbitrary ideal K of $\bar{A}$ its annihilator $Ann_{\bar{A}}K$ is also an ideal (see [Z1]).

Let $Q = \cup_{i\geq 0} Ann_{\bar{A}}K^{<i>}$. For an arbitrary derivation $d \in \mathcal{D}$, an arbitrary element $a \in \bar{A}$, if $q \in Ann_{\bar{A}}K^{<i>}$, then $d(q) \in Ann_{\bar{A}}K^{<i+1>}$ and $q \cdot a \in Ann_{\bar{A}}K^{<i+1>}$.

Indeed, $d(q) \cdot K^{<i+1>} \subseteq d(q \cdot K^{<i+1>}) + q \cdot d(K^{<i+1>}) = (0)$, because $d(K^{<i+1>}) \subseteq K^{<i>}$. Furthermore,

$$D(d(q), K^{<i+1>}) \subseteq [D(q, K^{<i+1>}), d] + D(q, d(K^{<i+1>})) = (0),$$

and

$$(q \cdot a)\dot{K}^{<i+1>} \subseteq q \cdot (a\dot{K}^{<i+1>}) + aD(q, K^{<i+1>}) = (0)$$

because $a \cdot K^{<i+1>}) = aR((K^{<i>})^3) \subseteq K^{<i>}$, by the Jordan identity.

Thus, Q is a graded $\mathcal{D}$-invariant ideal of $\bar{A}$ and $L \subseteq Q \cap P$.

Since $Q \cap P$ is a nonzero $\mathcal{D}$-invariant ideal of $\bar{A}$ lying in P, we have $(Q \cap P)_0 = P_0$, so $P_0 \subseteq Q$. Hence, there exists $i \geq 1$ such that $P_0 \subseteq Ann_{\bar{A}} K^{<i>}$.

Let $\tilde{L} = Ann_{\bar{A}} K^{<i>} \cap P$. Clearly $\tilde{L} \cdot K^{<i>} = (0)$ and we have shown that $P_0 \subseteq \tilde{L}$. In the same way as above, we can show that there exists $j \geq 1$ such that $P_0 \subseteq Ann_{\bar{A}} \tilde{L}^{<j>}$. Hence, $P_0^{<j+1>} = (0)$. The algebra P_0 is solvable and therefore nilpotent, the contradiction. Lemma is proved. □

LEMMA 1.4. *The algebra $\bar{A}$ is a direct sum of prime nondegenerate $\mathcal{D}$-invariant graded ideals*

PROOF. remark that a direct summand of $\bar{A}$ is automatically $\mathcal{D}$-invariant, since $\bar{A} = P' \oplus P''$ implies that $P'^2 = P'$.

Let $\bar{A} = \bar{A}^{(1)} \oplus \cdots \oplus \bar{A}^{(s)}$ be a maximal decomposition into a direct sum of nonzero graded ideals. We claim that each summand $\bar{A}^{(i)}$ is 0-minimal.

Indeed, let P be a 0-minimal $\mathcal{D}$-invariant graded ideal of $\bar{A}$, which is contained in $\bar{A}^{(i)}$. Suppose that $P \neq \bar{A}^{(i)}$.

The center $Z(\bar{A}^{(i)})$ of $\bar{A}^{(i)}$ lies in the center of $\bar{A}$, hence it does not contain nonzero nilpotent elements. Since $\bar{A}^{(i)}$ is indecomposable into a direct sum of two graded ideals it follows that $Z(\bar{A}^{(i)}) \cap \bar{A}_0^{(i)}$ is a field. This implies that $P_0 \cap Z(\bar{A}^{(i)}) = (0)$. Since $\bar{A}^{(i)}$ is nondegenerate we have $Z(P) \subseteq Z(\bar{A}^{(i)})$ (see [Z3]). Hence, $Z(P) \cap P_0 = (0)$.

From the classification in [MZ2] (see also the Introduction) it follows that the prime nondegenerate algebra P is one-sided graded. Suppose that P is "positively graded". Then $Z(P) \subseteq \sum_{i \geq 1}^{\infty} P_i$.

Let $P = \sum_{i \geq -l} P_i$. Clearly $PZ(P)^{l+1} \subseteq \sum_{i \geq 1} P_i$.

We will show that for an arbitrary $k \geq 1$ we have $PZ(P)^k \cdot \bar{A} \subseteq PZ(P)^{k-1}$. Indeed, the center $Z(P)$ is invariant with respect to all derivations of P. For arbitrary elements $z_1, \ldots, z_k \in Z(P)$; $a \in \bar{A}$ we have

$$PR(z_1) \cdots R(z_k)R(a) \quad \subseteq \quad PR(z_1) \cdots R(z_{k-1})R(a)R(z_k) \quad +$$
$$PR(z_1) \cdots R(z_{k-1})D(a, z_k) = PR(z_1) \cdots R(z_{k-1})R(a)R(z_k) \bmod PZ(P)^{k-1}.$$

Proceeding in this way we get

$$PR(z_1) \cdots R(z_k)R(a) = PR(a)R(z_1) \cdots R(z_k) \, mod \, PZ(P)^{k-1},$$

but

$$PR(a)R(z_1) \cdots R(z_k) \subseteq PZ(P)^k.$$

We have already mentioned in the Introduction that $R < \bar{A} >$ is the linear span of operators of the type $R(a)R(b)D_1 \cdots D_r$, where $a, b \in \bar{A}$; $D_i \in D(\bar{A}, \bar{A})$. Hence,

$$PZ(P)^{l+3}R < \bar{A} > \subseteq PZ(P)^{l+1} \subseteq \sum_{i \geq 1} P_i.$$

Now $PZ(P)^{l+3}R < \bar{A} >$ is a D-invariant ideal of $\bar{A}$ such that $(PZ(P)^{l+3}R < \bar{A} >)_0 = (0)$, the contradiction. Lemma is proved. □

Let $\bar{e}^{(i)}$ be the identity element of $\bar{A}^{(i)}$. Clearly $\bar{e}^{(i)} \in \bar{A}_0^{(i)}$. Let $\{\bar{e}_\mu^{(i)}\}_\mu$ be a maximal system of pairwise orthogonal idempotents of $\bar{A}_0^{(i)}$. So $\sum_\mu \bar{e}_\mu^{(i)} = \bar{e}^{(i)}$ and for each μ the subalgebra $\bar{A}_0^{(i)} U(\bar{e}_\mu^{(i)})$ does not contain proper idempotents.

The system $\{\bar{e}_\mu^{(i)}\}_{i,\mu}$ can be lifted to a system of pairwise orthogonal idempotents $e_\mu^{(i)} \in A_0$, such that $\sum_{i,\mu} e_\mu^{(i)} = 1$ (see [J3]).

Definition 1.5. An element $a \in J$ is said to be **Peirce homogeneous** if it lies in one of the subspaces $JU(e_\mu^{(i)}, e_\nu^{(j)})$, $(i, \mu) \neq (j, \nu)$ or in $\sum_{i,\mu} JU(e_\mu^{(i)})$.

Definition 1.6. An element $a \in J$ is said to be **strongly Peirce homogeneous** if it lies in one of the subspaces $JU(e_\mu^{(i)}, e_\nu^{(j)})$ or in one of the subspaces $JU(e_\mu^{(i)})$.

Define the trace functional $t : A \longrightarrow F$ via $t(A_i) = 0$ for $i \neq 0$, $t : A_0 \longrightarrow F$ is the reduced trace, that is $t(a) = \sum_{i,\mu} \alpha_{i,\mu}$, where $\alpha_{i,\mu} \in F$, $\bar{a}U(\bar{e}_\mu^{(i)}) = \alpha_{i,\mu}\bar{e}_\mu^{(i)}$, for $a \in A_0$ (see [J3]). In particular, $t(I) = 0$.

Recall that for elements $x_\alpha \in J_{\bar{\alpha}}$, $y_\beta \in J_{\bar{\beta}}$, $z_\gamma \in J_{\bar{\gamma}}$; $\alpha, \beta, \gamma \in \{0, 1\}$ the Jordan triple product is definitionined as $\{x_\alpha, y_\beta, z_\gamma\} = (x_\alpha y_\beta)z_\gamma + x_\alpha(y_\beta z_\gamma) - (-1)^{\beta\gamma}(x_\alpha z_\gamma)y_\beta$.

Let $N(A_0)$ denote the radical of the finite dimensional algebra A_0.

Lemma 1.7. *Let $a \in I + N(A_0)$, $d \in \mathcal{D}$, $x, y \in M$. Then*
(i) $t(ad) = 0$,
(ii) $t(\{a, x, y\}) = 0$,
(iii) $t([xU(a), y]) = 0$.

Proof. (i) If $a \in I$ then $ad \in I$ and therefore $t(ad) = 0$. Let $a \in N(A_0)$. Without loss of generality we can assume that the derivation d is homogeneous. If $t(ad) \neq 0$, then d has degree 0. But $N(A_0)$ is invariant under all derivations of degree 0. Hence $ad \in N(A_0)$ and $t(ad) = 0$.

(ii) We have $\{a, x, y\} = aD(x, y) + a\cdot[x, y]$. From (i) it follows that $t(aD(x, y)) = 0$.

(iii) It is not difficult to check that $[xU(a), y] = 2\{a, x \cdot a, y\} - \{a^2, x, y\}$ after which it remains to refer to (ii). Lemma is proved. □

The idea of the following Lemma is due to I. Kaplansky ([Kap]).

LEMMA 1.8. *Let x be a homogeneous Peirce homogeneous element from M and let y be a homogeneous strongly Peirce homogeneous element from M. Suppose that $t([x,y]) \neq 0$. Then there exists a scalar $0 \neq \psi \in F$ and an element $n \in N'(A_0) = \sum_{i,\mu} N(A_0)U(e_\mu^{(i)})$ such that*

$$[x,y] \cdot y = \psi y + y \cdot n$$

PROOF. From $t([x,y]) \neq 0$ it follows that $[x,y] \in \sum_{i,\mu} A_0 U(e_\mu^{(i)})$, so $[x,y] = \sum \alpha_{i\mu} e_\mu^{(i)} + n$, $n \in N'(A_0)$, $\alpha_{i\mu} \in F$.

If $y \in MU(e_\mu^{(i)})$ then $[x,y] = \alpha_{i\mu} e_\mu^{(i)} + n$, $\alpha_{i\mu} = t([x,y]) \neq 0$, and it remains to choose $\psi = \alpha_{i\mu}$.

If $y \in MU(e_\mu^{(i)}, e_\nu^{(j)})$ then $[x,y] = \alpha_{i\mu} e_\mu^{(i)} + \alpha_{j\nu} e_\nu^{(j)} + n$, $t([x,y]) = \alpha_{i\mu} + \alpha_{j\nu} \neq 0$. We have $y \cdot (\alpha_{i\mu} e_\mu^{(i)} + \alpha_{j\nu} e_\nu^{(j)}) = 1/2(\alpha_{i\mu} + \alpha_{j\nu})y$, so $\psi = 1/2(t([x,y]))$. Lemma is proved □

LEMMA 1.9. *Let x be a homogeneous strongly Peirce homogeneous element from M. Then for arbitrary elements $a_1, \ldots, a_r \in I$ we have $t([xR(a_1)\cdots R(a_r), x]) = 0$.*

PROOF. Without loss of generality we can assume that the elements $a_1, \ldots, a_r$ are homogeneous and strongly Peirce homogeneous. Let $w = R(a_1)\cdots R(a_r)$. The element xw is homogeneous and Peirce homogeneous (though we don't claim that xw is strongly Peirce homogeneous).

If $t([xw,x]) \neq 0$ then by Lemma 1.8 $[xw,x] \cdot x = xw' = \psi x + x \cdot n$, where $0 \neq \psi \in F$, $n \in N(A_0)$, $w' = [w, R(x)^2]$. This implies

$$xw'(\psi Id + R(n))^{-1} = x$$

Let $v = w'(\psi Id + R(n))^{-1}$, $xv = x$, $deg v = 0$. Consider the subalgebra B of A generated by the elements a_i, $a_i R(x)^2$, n, $1 \leq i \leq r$. Clearly, $B \subseteq I + N(A_0)$.

Let $R^M < B >$ be the subalgebra of $End_F M$ generated by linear transformations $R(b) : M \to M$, $m \to mb$. The subspace $U = U(B,B) + R^M < B > U(B,B)$ is a two sided ideal of $R^M < B >$. By Lemma 1.7 (iii) we have $t([M, MU]) = (0)$.

Consider the quotient algebra $R = R^M < B > /U$. definitionine on the vector space of R the new multiplication $u \star v = uv + vu$ and denote the resulting Jordan algebra as $R^{((+))}$. The mapping $B \longrightarrow R^{((+))}$, $b \to R(b) + U$ is a homomorphism of Jordan algebras. By Lemma 1.1 there exists $k \geq 1$ such that $(R^M < B >)_0^k \subseteq U$.

Since $v \in (R^M < B >)_0$ we have $v^k \in U$ and thus $x = xv^k \in MU$. This contradicts the assumption that $t([xw,x]) \neq 0$. Lemma is proved. □

Denote $e^{(i)} = \sum_\mu e_\mu^{(i)}$, $I_{\alpha ij} = \{e^{(i)}, I_\alpha, e^{(j)}\}$, $\alpha \in Z$.

Let Ω be the subalgebra of $R^M < A >$ generated by all operators from $R(N'(A_0))$, $D(I_{\alpha ij}, I_{-\alpha ij})$, $\alpha \in Z$; $i,j \geq 1$.

LEMMA 1.10. *Let x, y be homogeneous strongly Peirce homogeneous elements from M such that $[x, y] \in I$ and $[x, y\Omega] \subseteq I$. Then $t([x \cdot I, y]) = (0)$.*

PROOF. Choose a homogeneous strongly Peirce homogeneous element $a \in I$ and suppose that $t([x \cdot a, y]) \neq 0$. By Lemma 1.1.5 $[x \cdot a, y] \cdot y = \psi y + y \cdot n$, where $0 \neq \psi \in F$, $n \in N'(A_0)$. Let $b = [x, y]$, $a' = aR(y)^2$. Then $xR(y)^2 = yb$ and $(y \cdot b) \cdot a + x \cdot a' - y \cdot n = \psi y$.

We have $[xa, xa'] = [xR(a)R(a'), x] - a'D(xa, x)$. Hence $t([x \cdot a, x \cdot a']) = 0$ by Lemma 1.9 If $t([x \cdot a, y \cdot n]) \neq 0$ then we substitute $x := x$, $a := a$, $y := y \cdot n$ (remark that $y \cdot n$ is a homogeneous strongly Peirce homogeneous element).

If $t([x \cdot a, y \cdot n]) = 0$, then $t([x \cdot a, (y \cdot b) \cdot a]) \neq 0$.

We do not claim that the element $b \in I$ is strongly Peirce homogeneous. But b is a sum $b = \sum b'$ of homogeneous strongly Peirce homogeneous elements $b' \in I$. And for at least one of them we have $t([x \cdot a, (y \cdot b') \cdot a]) \neq 0$.

Recall that $R(b')R(a) = 1/2(U(b', a) + D(b', a) + R(b' \cdot a))$ and $t([x \cdot a, yU(b', a)]) = 0$ by Lemma 1.7(iii). If a homogeneous element of A has nonzero trace, then it lies in A_0. Since $t([xa, y]) \neq 0$ and $t([xa, (yb')a] \neq 0$ it follows that the sum of degrees of b' and a is zero. Hence $b'a \in N(A_0)$.

The Peirce decomposition of J with respect to $\{e_\mu^{(i)}\}_{i,\mu}$ defines a (Peirce) gradation of J by the abelian group $\oplus_{i,\mu}(\mathbb{Z}/2\mathbb{Z})x_\mu^{(i)}$ via $deg_P\{e_\mu^{(i)}, J, e_\nu^{(j)}\} = x_\mu^{(i)} - x_\nu^{(j)}$. If a Peirce homogeneous element of A has nonzero trace then its Peirce degree is zero. Arguing as above, we conclude that $deg_P(a) + deg_P(b') = 0$, hence $deg_P(ab') = 0$ which implies $ab' \in N'(A_0)$. If $t([x \cdot a, y \cdot (b' \cdot a)]) \neq 0$ then we can substitute $x := x$, $a := a$, $y := y \cdot (b' \cdot a)$.

Now it remains to consider the case when $t([x \cdot a, yD(b', a)]) \neq 0$. Since $yD(b', a)$ is a strongly Peirce homogeneous element, we can substitute $x := x$, $a := a$, $y := yD(b', a)$.

Let $a \in I_\alpha$. Let B be the subalgebra of A generated by I_α, $I_{-\alpha}$, $N'(A_0)$. By Lemma 1.1 the algebra $(R^M < B >)_0$ acts locally nilpotently on $M/\{B, M, B\}$. Let S be the subalgebra of $(R^M < B >)_0$ generated by the finite dimensional spaces $D(I_{-\alpha}, a)$ and $R(N'(A_0))$. It follows from Lemma 1.1 that there exists $r \geq 1$ such that $MS^r \subseteq \{B, M, B\}$. We have proved however (see the substitutions above) that if $t([x \cdot a, y]) \neq 0$ then for an arbitrary $k \geq 1$, $t([x \cdot a, yS^k]) \neq 0$, which contradicts Lemma 1.7(iii). Lemma is proved. □

Denote $I_{ij} = \{e^{(i)}, I, e^{(j)}\} = \sum_{\alpha \in \mathbb{Z}} I_{\alpha ij}$.

LEMMA 1.11. *$[I_{ij} \cdot M, M] \subseteq I$ for $i \neq j$.*

PROOF. Choose two arbitrary strongly Peirce homogeneous elements $x, y \in M$. If $a \in I_{ij}$ and $t([x \cdot a, y]) \neq 0$, then $[x, y]$ and $[x, y\Omega]$ both lie in $\{e^{(i)}, A, e^{(j)}\} \subseteq I$. By Lemma 1.10 we have $t([x \cdot I, y]) = (0)$, the contradiction. Hence, $t([I_{ij} \cdot M, M]) = 0$.

The sum $I_{ij}R(M)R(M) + I$ is an ideal in A. Indeed, choose arbitrary elements $a \in A$; $x, y \in M$. If $a \in I$ then $I_{ij}R(x)R(y)R(a) \subseteq I$. Otherwise we can assume that $a \in \cup_{k \geq 1} AU(e^{(k)})$. By the Jordan identity we have

$R(x)R(y)R(a) = R(a)R(y)R(x) - R([xa, y]) + R(xa)R(y) - R(ya)R(x) + R([x, y])R(a)$.

Considering summands on the right hand side we get $I_{ij}R(a) \subseteq I_{ij}$ and therefore $I_{ij}R(a)R(y)R(x) \subseteq I_{ij}R(M)R(M)$; the subspaces $I_{ij}R([xa, y])$ and $I_{ij}R([x, y])R(a)$ lie in I.

This proves that $I_{ij}R(x)R(y)R(a) \subseteq I_{ij}R(M)R(M) + I$.

Thus $I_{ij}R(M)R(M) + I \trianglelefteq A$.

Next we will notice that for arbitrary elements $a_1, \dots, a_r \in A$ we have

$$I_{ij}R(a_1)\cdots R(a_r)R(M)R(M) \subseteq I_{ij}R(M)R(M) + I.$$

Indeed, applying the Jordan identity to $R(a_r)R(M)R(M)$ we can use the induction assumption on r.

For an arbitrary derivation $d \in \mathcal{D}$ we have $I_{ij}d = \{e^{(i)}, I_{ij}, e^{(j)}\}d \subseteq \{e^{(i)}d, I_{ij}, e^{(j)}\} + \{e^{(i)}, I_{ij}d, e^{(j)}\} + \{e^{(i)}, I_{ij}, e^{(j)}d\} \subseteq I_{ij}R < A >$.

In view of the last two observations we conclude that the ideal $I_{ij}R(M)R(M) + I$ of A is $\mathcal{D}$-invariant.

From $t(I_{ij}R(M)R(M) + I) = (0)$ it follows that the zero component of $I_{ij}R(M)R(M) + I$ is nilpotent. Hence, from maximality of I it follows that $I_{ij}R(M)R(M) \subseteq I$. Lemma is proved. □

LEMMA 1.12. *Let x, y be homogeneous strongly Peirce homogeneous elements from M and let a be a homogeneous Peirce homogeneous element, $a \in I$, such that $t([x \cdot a, y]) \neq 0$. Then for an arbitrary $k \geq 1$ there exists a homogeneous strongly Peirce homogeneous element y_k such that $t([x \cdot a^{2^k}, y_k]) \neq 0$.*

PROOF. Since $N'(A_0)$ acts nilpotently (see [J3]) we can assume without loss of generality we can assume that $t([x \cdot a, y \cdot N'(A_0)]) = (0)$.

From Lemma 1.8 it follows that $t([x\cdot a, (y\cdot b)\cdot a]) \neq 0$ or $t([x\cdot a, y\cdot n]) \neq 0$, where $b = [x, y]$, $n \in N'(A_0)$ (see the proof of Lemma 1.10). In view of our assumption the second case is imposible. Hence, $t([x \cdot a, (y \cdot b) \cdot a] = t([(x \cdot a) \cdot a, y \cdot b]) = \frac{1}{2}t([xa^2, yb]) \neq 0$ by Lemma 1.7.

For some strongly Peirce homogeneous component y' of $y \cdot b$ we have $t([x \cdot a^2, y']) \neq 0$. And so on. Lemma is proved. □

LEMMA 1.13. *Let $\bar{A}^{(j)}$ be a one-sided graded algebra. Then for an arbitrary i we have $t([I_{ii} \cdot M_{ij}, M_{ij}]) = 0$.*

PROOF. Let $x, y \in M_{ij}$ be homogeneous strongly Peirce homogeneous elements and let $a \in I_{ii}$ be a homogeneous Peirce homogeneous element such that $t([x\cdot a, y]) \neq 0$. By Lemma 1.12 for any $k \geq 1$ there exists a strongly Peirce homogeneous element $y_k \in M_{ij}$ such that $t([x \cdot a^{2^k}, y_k]) \neq 0$.

Suppose that $\bar{A}^{(j)}$ is a positively graded algebra, that is, $\bar{A}^{(j)}_\mu = 0$ for $\mu \leq -s$.

If $deg(a) = 0$ then the element a is nilpotent. So there exists $k \geq 1$ such that $a^{2^k} = 0$.

Suppose that $deg(a) > 0$. Choose k such that $(2^k - 1)deg(a) > s$.

If $i = j$ then $t([x \cdot a^{2^k}, y_k]) \neq 0$ implies that $[x, y_k]$, $[x, y_k\Omega] \subseteq \{e^{(j)}, A, e^{(j)}\}$, and $deg[x \cdot a, y_k] = deg[x \cdot a, y_k\Omega] = -(2^k - 1)deg(a) < -s$.

Hence $[x, y_k]$, $[x, y_k\Omega] \subseteq I$. By Lemma 1.10 we have $t([x \cdot a^{2^k}, y_k]) = 0$, the contradiction.

Let $i \neq j$. From $t([xU(a, a^{2^k-1}), y_k]) = 0$ (see Lemma 1.7) it follows that $t([x \cdot a^{2^k}, y_k]) = 2t([(x \cdot a) \cdot a^{2^k-1}, y_k])$. We have $deg[x \cdot a, y_k] = deg[x \cdot a, y_k\Omega] = -(2^k - 1)deg(a) < -s$. The element $x \cdot a$ is strongly Peirce homogeneous because $i \neq j$. Moreover $\{a, x, y_k\} = aD(x, y_k) + a[x, y_k] \in I$ by Lemma 1.11.

Hence $[x \cdot a, y_k] = \{x, a, y_k\} - \{a, x, y_k\} = \{x, a, y_k\} \ mod\ I$.

But $\{x, a, y_k\} \in \{e^{(j)}, A, e^{(j)}\}$. Hence $[x \cdot a, y_k] \in I$ and similarly $[x \cdot a, y_k\Omega] \subseteq I$. Again by Lemma 1.10 $t([(x \cdot a) \cdot a^{2^k-1}, y_k]) = 0$, the contradiction.

Now suppose that $deg(a) < 0$.

In [MZ2] it was shown that a positively graded algebra $\bar{A}^{(j)}$ contains a central element c_m of positive degree m, such that $\bar{A}_k^{(j)} = \bar{A}_{k-m}^{(j)} \cdot c_m$ for a sufficiently big k. Hence, there exists $r \geq 1$ such that

$$\sum_{k \geq r} \bar{A}_k^{(j)} \subseteq (\sum_{k > s} \bar{A}_k^{(j)})^2$$

or, equivalently

$$\sum_{k \geq r} \{e^{(j)}, A_k, e^{(j)}\} \subseteq (\sum_{k > s} \{e^{(j)}, A_k, e^{(j)}\})^2 + I.$$

We claim that for arbitrary elements $b', b'' \in \sum_{k \geq s} \{e^{(j)}, A_k, e^{(j)}\}$, arbitrary $x, y \in M$ there holds $t(\{(b' \cdot b''), x, y\}) = 0$. Indeed, $\{(b' \cdot b''), x, y\} = (b' \cdot b'')D(x, y) + (b' \cdot b'') \cdot [x, y] = b' \cdot b''D(x, y) + b'D(x, y) \cdot b'' + (b' \cdot b'') \cdot [x, y]$.

Now it remains to prove that for arbitrary elements $c \in A$, $b \in \sum_{k > s} \{e^{(j)}, A_k, e^{(j)}\}$, we have $t(b \cdot c) = 0$. We have $A = \sum_i \{e^{(i)}, A, e^{(i)}\} + I$. If $c \in I$ then $b \cdot c \in I$ and $t(b \cdot c) = 0$. If $c \in \{e^{(i)}, A, e^{(i)}\}$, $i \neq j$, then $b \cdot c = 0$. If $c \in \{e^{(j)}, A, e^{(j)}\}$ and c is homogeneous, then the only case when $t(b \cdot c)$ can be $\neq 0$ is when $deg(c) = -deg(b) < -s$. This implies $c \in I$. The claim is proved.

Now it follows that for arbitrary elements $x, y \in M$; $b \in \sum_{k \geq r} \{e^{(j)}, A_k, e^{(j)}\}$ we have $t(\{b, x, y\}) = 0$.

Consider the subalgebra $B = I + \sum_{k \geq r} \{e^{(j)}, A_k, e^{(j)}\} + N'(A_0)$ of A.

We have $t(\{B, M, M\}) = (0)$, which implies also $t([M, MU(B)]) = (0)$.

Choose $k \geq 1$ such that $l = |deg(a)|(2^k - 1) \geq r$. If x, y_k are homogeneous strongly Peirce homogeneous elements such that $t([x \cdot a^{2^k}, y_k]) \neq 0$, then $t([(x \cdot a) \cdot a^{2^k-1}, y_k]) \neq 0$ and $deg[x \cdot a, y_k] = l$.

Consider the subalgebra Ω' of Ω generated by $D(B_{-l}, B_l), R(N'(A_0))$. By Lemma 1.1 Ω' acts nilpotently on $M/MU(B)$. Hence, there exists $q \geq 1$ such that $t([M, M\Omega'^q]) = 0$. Hence there exists a homogeneous strongly Peirce homogeneous element y such that $t([(x \cdot a) \cdot a^{2^k-1}, y]) \neq 0$ but $t([(x \cdot a) \cdot a^{2^k-1}, y\Omega']) = 0$. Arguing as at the beginning of the proof of Lemma 1.10, we conclude that $t([x \cdot a) \cdot a^{2^k-1}, yD(b, a^{2^k-1})]) \neq 0$, where $b = [x \cdot a, y]$. Since $a \in I_{ii}$ and $x, y \in M_{ij}$ it follows that $\{x, a, y\} \in \{e^{(j)}, A, e^{(j)}\}$. Moreover $deg\{x, a, y\} = -deg(a)(2^k - 1) = l \geq r$. Since $b = [xa, y] = 1/2(\{x, a, y\} + \{a, x, y\})$

and $\{a, x, y\} \in I$, we get that $b \in B$ and therefore $D(b, a^{2^k-1}) \in \Omega'$. This contradicts our assumptions. Lemma is proved. □

LEMMA 1.14. *Let* $dim\bar{A}^{(j)} < \infty$. *Then for an arbitrary* i *we have* $t([I_{ii} \cdot M_{ij}, M_{ij}]) = 0$

PROOF. Just repeats the proof of Lemma 1.13. □

Consider the ideal $\tilde{I} = I + \sum\{e^{(j)}, A, e^{(j)}\}$, where the summation on the right hand side is taken over indices j that $\bar{A}^{(j)}$ is either a loop algebra or an infinite dimensional Jordan algebra of a bilinear form (see [MZ2]).

LEMMA 1.15. $IR(M)R(M) \subseteq \tilde{I}$.

PROOF. Suppose that $(IR(M)R(M) + I/I) \cap \bar{A}^{(j)} \neq (0)$, where $\bar{A}^{(j)}$ is finite dimensional or one-sided graded.

The ideal $I + AU(e^{(j)})$ is $\mathcal{D}$-invariant. Hence $(IR(M)R(M)+I) \cap (I+AU(e^{(j)}))$ is $\mathcal{D}$-invariant and strictly bigger than I. Hence the 0-component of this intersection is not nilpotent and thus has nonzero trace. This means that $t((IR(M)R(M) + I) \cap A_0U(e^{(j)})) \neq 0$.

Let $a_0U(e^{(j)}) \in \sum_\alpha I_{\alpha\alpha}R(M)R(M) + I$. Applying $U(e^{(j)})$ to both sides we obtain $a_0U(e^{(j)}) \in \sum_\alpha I_{\alpha\alpha}R(M)R(M)U(e^{(j)}) + I_{jj}$. By the Jordan identity we have $R(x)R(y)R(e^{(j)}) = R(e^{(j)})R(y)R(x) - R([x \cdot e^{(j)}, y])+$
$R(x \cdot e^{(j)})R(y) - R(y \cdot e^{(j)})R(x) + R([x, y])R(e^{(j)})$, for arbitrary elements $x, y \in M$.

This implies that

$$IR(M)R(M)U(e^{(j)}) \subseteq I_{jj}R(M)R(M) + \sum_{\alpha \neq j} I_{\alpha\alpha}R(M \cdot e^{(j)})R(M) + I.$$

Intersecting the right hand side with $\sum_i AU(e^{(i)})$ we get

$$IR(M)R(M)U(e^{(j)}) \subseteq \sum_i I_{jj}R(M_{ij})R(M_{ij}) + \sum_{\alpha \neq j} I_{\alpha\alpha}R(M_{\alpha j})R(M_{\alpha j}) + I.$$

For $i \neq j$ we have
$I_{jj} + I_{jj}R(M_{ij})R(M_{ij}) = \{I_{jj}, M_{ij}, M_{ij}\} + \{M_{ij}, I_{jj}, M_{ij}\} + I_{jj}$,
$\{I_{jj}, M_{ij}, M_{ij}\} \subseteq AU(e^{(j)})$, $\{M_{ij}, I_{jj}, M_{ij}\} \subseteq AU(e^{(i)})$. This implies the more precise inclusion
$IR(M)R(M)U(e^{(j)}) \subseteq I_{jj}R(M_{jj})R(M_{jj}) + \sum_{i \neq j}\{I_{jj}, M_{ij}, M_{ij}\} +$
$\sum_{\alpha \neq j}\{M_{\alpha j}, I_{\alpha\alpha}, M_{\alpha j}\} + I_{jj}$.
The right hand side has zero trace by Lemmas 1.13 and 1.14. Lemma is proved. □

LEMMA 1.16. *If* $i \neq j$ *then* $id_J(I_{ij}) \cap A \subseteq \tilde{I}$.

PROOF. Let $u_1, \ldots, u_l \in A \cup M$ be homogeneous strongly Peirce homogeneous elements, $w = R(u_1) \cdots R(u_l)$. Suppose further that $I_{ij}w \subseteq A$, $I_{ij}w \not\subseteq \tilde{I}$ and l is the smallest number with this property. Then

(1) there are no two consecutive elements u_k, u_{k+1} lying in A.

Indeed, if two consecutive elements u_k, u_{k+1} lie in A, then using the Jordan identity we can move them to the right end of w modulo operators of length $< l$.

(2) There are no two consecutive elements u_k, u_{k+1} lying in M.

Indeed, suppose that $u_k, u_{k+1} \in M$. If there is another odd element u_i among $u_1, \ldots, u_l$ then again using the Jordan identity we can represent w as a linear combination of operators $\cdots R(u_k)R(u_{k+1})R(u_i)\cdots$, $\cdots R(u_{k+1})R(u_k)R(u_i)\cdots$ of length l and operators of length $< l$. The identity (D) from the Introduction implies that for odd elements $x, y, z \in M$ the operator $R(x)R(y)R(z)$ is a linear combination of operators of length ≤ 2 and of operators of the type $R(x)D(y,z)$. We can move $D(y,z)$ to the right end and it remains to notice that $\tilde{I}D(M,M) \subseteq \tilde{I}$.

Suppose now that u_k, u_{k+1} are the only odd elements among $u_1, \ldots u_l$. Because of the Jordan identity we can assume $k = 1$.

Then $I_{ij}w \in I_{ij}R(M)R(M)R(A) \subseteq \tilde{I}$.

(3) No element u_k lies in I.

Indeed, moving $R(u_k)$ to the right via the Jordan identity we get $w = (\cdots)R(u_k) + (\cdots)R(u_k)R(v)$, $v \in A \cup M$ modulo operators on length $< l$. If $u_k \in I$ then $Aw \subseteq \tilde{I}$ by Lemma 1.1.12.

(4) For an element $c \in I_{ij}$, the expression $cw + \tilde{I}/\tilde{I}$ is skew symmetric with respect to all elements u_k lying in A and symmetric with respect to all elements u_k lying in M.

Since even (odd) elements do not follow one after another the assertion follows from the Jordan identity and minimality of l.

Let us finish the proof of the Lemma

Clearly $l \geq 3$. If $u_1 \in A$ then we can assume that $u_1 \in AU(e^{(\alpha)}, e^{(\beta)})$. If $\alpha \neq \beta$ then $u_1 \in I$ and we have shown that this is impossible. So, the only option is $u_1 \in AU(e^{(i)})$ or $u_1 \in AU(e^{(j)})$. But then $I_{ij}R(u_1) \subseteq I_{ij}$. Hence $u_1 \in M$ and consequently $u_2 \in A$. Since $u_2 \notin I$ we can assume that $u_2 \in AU(e^{(\alpha)})$. If $\alpha \neq i$, $\alpha \neq j$, then $D(I_{ij}, u_2) = 0$ and $I_{ij}R(u_1)R(u_2) = I_{ij}R(u_1u_2))$. Let $\alpha = j$. We have

$$I_{ij}w = I_{ij}R(e^{(i)})R(u_1)R(u_2)\cdots \subseteq I_{ij}(-R(u_2)R(u_1)R(e^{(i)})+$$

$$R(e^{(i)})R(u_1u_2) + R(u_2)R(u_1e^{(i)}))\cdots + \tilde{I} = I_{ij}R(u_2)R(u_1)R(e^{(i)})\cdots + \tilde{I}$$

Thus, without loss of generality we can assume that $u_2 = e^{(i)}$.

Let $u_1 \in \{e^{(k)}, M, e^{(i)} + e^{(j)}\}$, where $k \neq i$, $k \neq j$. Then $u_1U(I_{ij}, u_2) = (0)$, which implies $I_{ij}R(u_1)R(u_2) \subseteq I_{ij}R(u_2)R(u_1) + I_{ij}R(u_1u_2)$. This contradicts our earlier results.

If $u_1 \in MU(e^{(i)}) + MU(e^{(j)})$ then $R(u_1)$ commutes with $R(u_2) = R(e^{(i)})$.

The only remaining case is $u_1 \in M_{ij}$.

Since $I_{ij}w + \tilde{I}/\tilde{I}$ is symmetric in u_1 and u_3 we can assume also that $u_3 \in M_{ij}$. But then $I_{ij}R(u_1)R(u_2)R(u_3) \subseteq \{e^{(i)}, A, e^{(j)}\} = I_{ij}$.

Lemma is proved □

Since the superalgebra J is simple, Lemma 1.16 implies that either $\tilde{I} = A$ and so $\bar{A}$ is a direct sum of loop algebras and of infinite dimensional Jordan algebras of bilinear forms or $I_{ij} = (0)$ for $i \neq j$ and so $A = A^{(1)} \oplus \cdots \oplus A^{(r)}$ where each $A^{(i)}/I^{(i)}$ is a prime graded algebra.

Let's analize the case when $A = \tilde{I}$, that is, when A/I is a sum of loop algebras and of infinite dimensional Jordan algebras of bilinear forms.

We claim that in this case $I = M(A)$, the McCrimmon radical of A. Indeed, if a homogeneous element $a \in I$ is not nilpotent then there exists a graded ideal P of A such that a is not nilpotent modulo P and A/P is a prime nondegenerate algebra (it is sufficient to take a maximal graded ideal P of A such that a is not nilpotent modulo P, see [ZSSS]). Then
$A/(I+P) \simeq (A/P)/(I+P)/P$. From the classification theorem (see [MZ2]and the Introduction) it follows that either A/P is graded simple and $1 \in (A/P)_0$ or A/P is a one-sided graded algebra. Let A/P be graded simple. Since $I + P/P \neq (0)$ it follows that $A = I + P$.

Hence $(A/P)_0 = (I+P/P)_0$ is nilpotent, the contradiction. Suppose that A/P is a one sided graded algebra. We will show that in this case $A = I + P$ as well. Otherwise $A/I + P \simeq (A/I)/(I + P)/I$ is a sum of loop algebras and of infinite dimensional Jordan algebras of bilinear forms and thus can not be a homomorphic image of a one-sided graded algebra.

We showed that every homogeneous element of I is nilpotent modulo P. If $I/M(A) \neq (0)$ then $I/M(A)$ is a subdirect product of prime nondegenerate graded Jordan algebras (see [Z4]). Now it remains to notice that no algebra from the classification list of [MZ2] has all homogeneous elements nilpotent.

We proved that $I = M(A)$.

LEMMA 1.17. *If $A = \tilde{I}$ then $IR(M)R(M) \subseteq I$.*

PROOF. From the Jordan identity it follows that $IR(M)R(M) + I$ is an ideal in A. Since I was shown to be a nil ideal (under the assumption that $A = \tilde{I}$) it follows from Lemma 1.12 that $t(IR(M)R(M) + I) = (0)$. Now, by maximality of I we have $IR(M)R(M) \subseteq I$. Lemma is proved. □

LEMMA 1.18. *If $A = \tilde{I}$ and $i \neq j$ then $I_{ij} = 0$.*

The proof of this Lemma just verbatim follows the proof of Lemma 1.16 with the only change: to prove the inclusion $id_J(I_{ij}) \cap A \subseteq I$ we refer to Lemma 1.17 instead of Lemma 1.15.

Thus, if $A = \tilde{I}$ then $A = A^{(1)} \oplus \cdots \oplus A^{(r)}$, and each $A^{(i)}/M(A^{(i)})$ is either a loop algebra or an infinite dimensional Jordan algebra of a bilinear form.

LEMMA 1.19. *If* $\tilde{I} = A$, $I \neq (0)$ *then* $r = 1$.

PROOF. Since $I \neq (0)$ it follows that there exists i such that $I_{ii} = M(A^{(i)}) \neq 0$.

We will show that $id_J(I_{ii}) \cap A \subseteq A^{(i)} + I$.

Indeed, suppose that the assertion is not true. Choose an operator $w = R(u_1) \cdots R(u_l)$, $u_i \in A \cup M$, such that $I_{ii}w \subseteq A$, $I_{ii}w \not\subseteq A^{(i)} + I$ and l is the smallest with this property. Then $l \geq 3$ since $I_{ii}R(M)R(M) \subseteq I$ by Lemma 1.17. Arguing as in the proof of Lemma 1.16 we see that no two consecutive elements u_k, u_{k+1} lie in A (resp. in M). The element u_1 clearly lies in M. Hence $u_2 \in A$.

If $u_2 \in A^{(j)}$, $j \neq i$ then $R(I_{ii})$ and $R(u_2)$ commute, so $I_{ii}R(u_1)R(u_2) = I_{ii}R(u_1 \cdot u_2)$.

Hence we can assume that $u_2 \in A^{(i)}$. If $u_1 \in M_{ik}$, $k \neq i$, then $(0) = u_1 U(I_{ii}, u_2)$ and therefore $I_{ii}R(u_1)R(u_2) \subseteq I_{ii}R(u_1u_2) + I_{ii}R(u_2)R(u_1)$.

Hence we can assume that $u_1 \in M_{ii}$. Since for an arbitrary element $c \in I_{ii}$ the expression $cw + A^{(i)} + I/A^{(i)} + I$ is symmetric in odd $u_k's$ and skew-symmetric in even $u_k's$ we can assume that $u_1, \ldots, u_l \in A^{(i)} \cap M_{ii}$, the contradiction. Lemma is proved. □

Now suppose that $\tilde{I} \neq A$, but still $I \neq (0)$. Since $\tilde{I} \neq A$ it follows that $A = A^{(1)} \oplus \cdots \oplus A^{(r)}$ and for at least one i the algebra $\bar{A}^{(i)}$ is either finite dimensional or one-sided graded.

The following Lemma was proved in [RZ].

LEMMA 1.20. *Let J be a graded simple unital Jordan superalgebra, $J = A + M$. Then A can not be represented as a direct sum of 3 nonzero ideals.*

Hence, if $r \geq 2$ then $A = A^{(1)} \oplus A^{(2)}$. Suppose that $\bar{A}^{(2)}$ is finite dimensional or one-sided graded.

LEMMA 1.21. *For any* $1 \leq i \leq r$ *we have*

$$id_J(I_{ii}) \cap A \subseteq \tilde{I} + \{e^{(i)} A e^{(i)}\} \trianglelefteq A.$$

PROOF. Again suppose the contrary and choose an operator $w = R(u_1) \cdots R(u_l)$ such that $u_i \in A \cup M$, $I_{ii}w \subseteq A$, $I_{ii}w \not\subseteq \tilde{I} + A^{(i)}$ and l is the smallest number with this property. Then, by Lemma 1.15 we have $l \geq 3$. Arguing as in the proof of Lemma 1.19 we see that (1) there are no two consecutive even (odd) elements among $u_1, \ldots, u_l$; (2) all elements u_k, $1 \leq k \leq l$ lie in $A^{(i)} + M_{ii}$. But then $I_{ii}w \subseteq A^{(i)}$. Lemma is proved. □

COROLLARY 1.22. $A^{(1)}$ is a loop algebra or an infinite dimensional Jordan algebra of a bilinear form

Indeed, if $\bar{A}^{(1)}$ is also one side graded or finite dimensional, then $\tilde{I} = I$. From Lemma 1.21 and simplicity of J it follows that $I_{11} = I_{22} = (0)$, so $I = (0)$, which contradicts our assumption. Hence $\bar{A}^{(1)}$ is a loop algebra or an infinite dimensional algebra of a bilinear form.

If $I_{11} \neq (0)$ then $id_J(I_{11}) \cap A \subseteq \tilde{I} + A^{(1)} = I + A^{(1)} \neq A$, the contradiction. This proves the corollary

LEMMA 1.23. $M = M_{12}$

PROOF. If $M_{12} = 0$, then $J = (A^{(1)} + M_{11}) \oplus (A^{(2)} + M_{22})$.

Assume therefore that $M_{12} \neq (0)$.

Consider the ideal $Ann_{12} = \{a \in A^{(1)} | a \cdot M_{12} = 0\}$ of $A^{(1)}$.

Since the algebra $A^{(1)}$ is graded simple and $e^{(1)} \notin Ann_{12}$ it follows that $Ann_{12} = (0)$

We have already shown in the proof of Lemma 1.19 that $[M_{11}, M_{12}] = \{M_{11}, M_{12}, M_{11}\} = (0)$ implies $[M_{11}, M_{11}] \subseteq Ann_{12} = 0$. Hence M_{11} is an ideal of J and therefore $M_{11} = (0)$. Again as in the proof of Lemma 1.19 we can show that $Ann_{21} = (0)$ (remark that in this case $I_{22} \neq (0)$ as in Lemma 1.19), which implies $M_{22} = (0)$. Lemma is proved. □

LEMMA 1.24. $id_J(I_{22}) \cap A \subseteq A^{(1)} + I_{22} = \tilde{I}$

PROOF. Suppose the contrary. As before, let $u_1, \ldots, u_l \in A \cup M$, $w = R(u_1) \cdots R(u_l)$, $I_{22}w \subseteq A$, $I_{22}w \nsubseteq A^{(1)} + I_{22}$ and l minimal with this property. From Lemma 1.15 it follows that $l \geq 3$. As many times before we conclude that there are no consecutive even (odd) elements among $u_1, \ldots, u_l$ and that $u_1, u_3 \in M$, $u_2 \in A$.

If $u_2 \in A^{(1)}$ then $D(I_{22}, u_2) = (0)$ and therefore $I_{22}R(u_1)R(u_2) \subseteq I_{22}R(u_1 \cdot u_2) + I_{22}R(u_2)R(u_1)$. Let $u_2 \in A^{(2)}$. Since $M = M_{12}$ it follows that $\{I_{22}, u_1, u_2\} = (0)$. Hence, again $I_{22}R(u_1)R(u_2) \subseteq I_{22}R(u_1 \cdot u_2) + I_{22}R(u_2)R(u_1) \subseteq I_{22}R(J)$, the contradiction. Lemma is proved. □

COROLLARY 1.25. $I_{22} = (0)$. In other words, if $\tilde{I} \neq A$, then either A is a prime nondegenerate algebra or $A = A^{(1)} \oplus A^{(2)}$, where $A^{(1)}$, $A^{(2)}$ are prime nondegenerate algebras.

The following Proposition summarizes what we have proved so far about the structure of A.

PROPOSITION 1.26. Let $J = A + M$ be a graded simple Jordan superalgebra with dimensions $\dim J_i$ uniformely bounded. Let I be the maximal $\mathcal{D}$-invariant graded ideal of A such that I_0 is nilpotent. Then one of the following statements holds:

(1) $I = M(A) \neq (0)$, $A/M(A)$ is a loop algebra, or

(2) $I = M(A) \neq (0)$, $A/M(A)$ is an infinite dimensional Jordan algebra of a bilinear form, or

(3) A/I is finite dimensional or one-sided graded (including the case when $I = (0)$), or

(4) $I = (0)$, $A = A^{(1)} \oplus A^{(2)}$, where $A^{(i)}$ are prime nondegenerate algebras, or

(5) A is a loop algebra, or
(6) A is an infinite dimensional simple Jordan algebra of a bilinear form.

LEMMA 1.27. $[M \cdot I, M] \subseteq I$

PROOF. In cases (1) and (2) of Proposition 1.26 the assertion follows from Lemma 1.17. In case (3) the assertion follows from Lemma 1.15. Lemma is proved. □

LEMMA 1.28. $t([(M \cdot I) \cdot A, M]) = 0$

PROOF. Let's show that for an arbitrary element $b \in A$ we have $D(I,b)D(I,b) \subseteq R < A > R(I)$.

Indeed, let $a', a'' \in I$. Then

$$D(a',b)D(a'',b) = (R(a')R(b) - R(b)R(a'))R(a'')R(b) \; mod \, R < A > R(I).$$

The Jordan identity implies that $R(I)R(I)R(b) \subseteq R < A > R(I)$.

Hence, $R(b)R(a')R(a'')R(b) \in R < A > R(I)$.

And finally $R(a')R(b)R(a'')R(b) = \frac{1}{2}R(a')R(a'')R(b^2) = 0 \text{ mod } R < A > R(I)$ again by the Jordan identity.

Next we will show that for arbitrary elements $a \in I$; $b, b' \in A$, arbitrary derivation $d \in D$ we have

$$D(a,b)D(a,b')[D(a,b')D(a,b), d] \in R < A > R(I).$$

Indeed, from what we have proved above it follows that

$$D(a,b')D(a,b') \in R < A > R(I).$$

Hence

$$D(a,b)D(a,b')D(a,b')[D(a,b),d] \in R < A > R(I)D(A,A) \subseteq R < A > R(I).$$

Denote $[D(a,b'),d] = \tilde{D}$. Then

$$D(a,b)D(a,b')\tilde{D}D(a,b) =$$

$$D(a,b')D(a,b)\tilde{D}D(a,b) + (D(aD(a,b'),b) + D(a,bD(a,b'))\tilde{D}D(a,b) \in$$
$$D(a,b')D(a,b)\tilde{D}D(a,b) + (D(I,b) + D(I,I))\tilde{D}D(a,b).$$

Analizing summands on the right hand side we get

$$D(I,b)\tilde{D}D(a,b) \subseteq \tilde{D}D(I,b)D(a,b) + (D(I\tilde{D},b) + D(I,b\tilde{D}))D(a,b) \subseteq$$
$$R < A > D(I,b)D(I,b) + D(I,I)D(a,b) \subseteq R < A > R(I).$$

This implies also that $D(a,b')D(a,b)\tilde{D}D(a,b) \in R < A > R(I)$. The inclusion $D(I,I)\tilde{D}D(a,b) \subseteq R < A > R(I)$ is clear.

Now let's finish the proof of the Lemma. Choose homogeneous strongly Peirce homogeneous elements $x, y \in M$; $a \in I$, $b \in A$ and suppose that $t([(x \cdot a) \cdot b, y]) \neq 0$. Without loss of generality we will assume that
$t([(x \cdot N'(A_0))R(I)R(A), M]) = (0)$ and $t([xR(I)R(A), y \cdot N'(A_0)]) = (0)$. We have $t([(x \cdot a) \cdot b, y]) = t([xD(a,b), y])$. Let $' = R(y)^2$. By Lemma 1.8
$(x'D(a,b) + xD(a',b) + xD(a,b')) = \xi y + yn$ for some $0 \neq \xi \in F$, $n \in N'(A_0)$, so y

is a linear combination of $x'D(a,b)$, $xD(a',b)$, $xD(a,b')$, yn. From the assumption about y it follows that $t([xD(a,b), y\cdot n]) = 0$.

Since $D(a,b)D(a,b)$ and $D(a,b)D(a',b)$ lie in $R<A>R(I)$, we conclude that $t([xD(a,b), x'D(a,b)]) = -t([xD(a,b)D(a,b), x']) = 0$ and $t([xD(a,b), xD(a',b)]) = -t([xD(a,b)D(a',b), x]) = 0$.
Hence, $t([xD(a,b)D(a,b'), x]) \neq 0$.

Again by Lemma 1.8 x can be represented as a linear combination of $xD(a,b)D(a,b')R(x)^2$ and of xn for some element $n \in N'(A_0)$.
We have $t([xD(a,b)D(a,b'), xn]) = -t([xD(a,b), xR(n)D(a,b')]) = 0$ because $t([M, xR(n)R(I)R(A)]) = (0)$ by the assumption.

Denote $d = R(x)^2$. Then $xD(a,b)D(a,b')d = x[D(a,b)D(a,b'), d]$ and it is easy to see that

$$t([xD(a,b)D(a,b'), x[D(a,b)D(a,b'), d]) = t([xD(a,b)D(a,b')[D(a,b')D(a,b), d], x]).$$

In view of what we proved above the last expression is zero. Lemma is proved.

□

For elements $x, y \in M$; $u \in I$; $a, b \in A$ denote

$$t(x,y,u,a,b) = t([(x\cdot u)\cdot a, y]\cdot b).$$

LEMMA 1.29. *The multilinear function $t : M \times M \times I \times A \times A \longrightarrow F$ has the following properties:*
(1) $t(x,y,u,a,b) = -t(x,y,u,b,a)$,
(2) $t(x,y,u,a,b) = t(y,x,u,a,b)$,
(3) $t(x,y,u,a^2,b) = 2t(xa,y,u,a,b) = 2t(x,ya,u,a,b) = 2t(x,y,ua,a,b) = 2t(x,y,u,a,ab)$.

PROOF. (1) By the Jordan identity $2[(xu)a,y]a = -[(xu), ya^2] + 2[(xu)a, ya] + [xu,y]a^2$.
Hence, $t(x,y,u,a,a) = t([(xu)a,y]a) = 0$ by Lemmas 1.27, 1.28.

(2) $[(xu)a,y] = [xD(u,a),y]\,(mod\,I) = -[x, yD(u,a)]\,(mod\,I) = [yD(u,a),x] = [(yu)a,x]\,(mod\,I)$.

(3) We have $[xR(u)R(a^2),y] = [xD(u,a^2),y]\,(mod\,I) = 2[xD(ua,a),y] = 2[(x(ua))a,y]\,(mod\,I)$.
Also, $(xu)R(a^2)R(y) = (xu)D(a^2,y)(mod\,I) = 2(xu)D(a,ya) = 2[(xu)a,ya](mod\,I)$.

In view of (2) we can conclude that $t(x,y,u,a^2,b) = 2t(xa,y,u,a,b)$.
Finally, $t(x,y,u,a^2,b) = 2t(x,y,u,a,ab)$ is equivalent to $t(x,y,u,a^2,a) = 0$. We have $t(x,y,u,a^2,a) = 2t(xa,y,u,a,a) = 0$ by (1). Lemma is proved. □

LEMMA 1.30. *For arbitrary elements $x, y \in M$, $u \in I$, $a, b, c, e \in A$ we have:*
(1) $t(x,y,u,cD(a,b),e) = \sum_\alpha t(x_\alpha, y_\alpha, u, a, e) + \sum_\beta t(x_\beta, y_\beta, u, b, e)$ *for some elements* $x_\alpha, y_\alpha, x_\beta, y_\beta \in M$.
(2) $t(x,y,uD(a,b),a,b) = 0$.

PROOF. (1) By Lemma 1.29 (3) modulo expressions of the right hand side type we have

$$t(x,y,u,(cb)a,e) = t(x,ya,u,cb,e) = t(xb,ya,u,c,e) = \\ t(xb,y,u,ca,e) = t(x,y,u,(ca)b,e).$$

This proves (1).

(2) $t(x,y,(ua)b,a,b) = 1/2t(x,y,ua,a,b^2) = 1/4t(x,y,u,a^2,b^2) = t(x,y,(ub)a,a,b)$. Lemma is proved. □

Let FJ denote the free Jordan algebra on the countable set of free generators $x_1, x_2, \cdots$ (see [J3], [ZSSS]).

As always, by $FJD(x_1,x_2)$ we denote the image of the derivation $D(x_1,x_2) : FJ \longrightarrow FJ$ and by $< FJD(x_1,x_2) >$ the subalgebra of FJ generated by $FJD(x_1,x_2)$.

Let $f(x_1,x_2,x_3) = (x_3D(x_1,x_2)^2)^2D(x_1,x_2)$. In [Z3] it was proved that for any $i \geq 1$ we have $x_4 \cdot f^i \in < FJD(x_1,x_2) >$.

LEMMA 1.31. *Let* $f = f(a,b,c)$; $a,b,c \in A$. *Then*
(1) $t(M,M,I,f^4,A) = (0)$,
for an arbitrary element $e \in A$ *we have*
(2) $t(M,M,I,f^{12}e,A) = (0)$,
(3) $t(M,M,I,(f^{24}e)e,A) = (0)$.

PROOF. Let $AD(a,b)$ be the image of the derivation $D(a,b) : A \to A$. Let $A_{a,b}$ be the subalgebra of A generated by $AD(a,b)$. By Lemmas 1.29 (3) and 1.30 (1), for arbitrary elements $x,y \in M$; $u \in I$; $c,c',c'' \in A_{a,b}$ we have:

$$t(x,y,u,c,b) = \sum t(\cdots,\cdots,u,a,b), \tag{1.1}$$

$$t(x,y,u,c',c'') = \sum t(\cdots,\cdots,u,a,b). \tag{1.2}$$

Let us show that

$$t(x,y,uc,a,b) = \sum t(\cdots,\cdots,u,a,b). \tag{1.3}$$

Indeed, $t(x,y,uc,a,b) = t(x,y,u,ca,b) - t(x,y,ua,c,b)$.
Furhermore, $t(x,y,u,ca,b) = t(xc,y,u,a,b) + t(xa,y,u,c,b)$ and it remains to apply (1.1) to $t(xa,y,u,c,b)$. Again by (1.1), $t(x,y,ua,c,b) = \sum_\alpha t(x_\alpha,y_\alpha,ua,a,b)$.

But $t(x_\alpha,y_\alpha,ua,a,b) = \frac{1}{2}t(x_\alpha,y_\alpha,u,a^2,b) = t(x_\alpha a,y_\alpha,u,a,b)$ which finishes the proof of (1.3). Let us show that an arbitrary element from If^2 is a linear combination of elements of the type $uD(a,b)R(c_1)\cdots R(c_r)$, where $u \in I$; $c_1,\ldots,c_r \in A_{a,b}$. Indeed, the element $x_4f^2(x_1,x_2,x_3)$ is a linear combination of products of elements of the type $gD(x_1,x_2)$, where g is a homogeneous element in x_1,x_2,x_3,x_4. Each product of this linear combination has degree 1 in x_4. This implies the claim.

Now we are ready to prove the first assertion of the Lemma. By Lemma 1.29 we have $t(M, M, I, f^4, A) \subseteq t(M, M, If^2, f, Af)$.

Since $Af \subseteq A_{a,b}$ and an arbitrary element of If^2 is a linear combination of elements of the type $uD(a,b)R(c_1)\cdots R(c_r)$; $u \in I$; $c_i \in A_{a,b}$ it is sufficient to prove that $t(M, M, uD(a,b)R(c_1)\cdots R(c_r), A_{a,b}, A_{a,b}) = (0)$.

By (1.2) for arbitrary elements $c', c'' \in A_{a,b}$ we have
$t(\cdots, \cdots, uD(a,b)R(c_1)\cdots R(c_r), c', c'') =$
$\sum t(\cdots, \cdots, uD(a,b)R(c_1)\cdots R(c_r), a, b)$ and by (1.3)
$t(\cdots, \cdots, uD(a,b)R(c_1)\cdots R(c_r), a, b) = \sum t(\cdots, \cdots, uD(a,b), a, b) = 0.$

(2) We have $f^{12}e = (f^8e)f^4 + f^8D(f^4, e)$. For a homogeneous polynomial h, by h(A) we denote the set of values of h on A. Since the ground field F is infinite it follows (see [Z6]) that the linear span $Fh(A)$ is invariant with respect to all derivations of A. In particular $f^8D(f^4, e) \in Ff^8(A)$ and therefore $t(M, M, I, f^8D(f^4, e), A) = 0$.

Furthermore, $t(M, M, I, (f^8e)f^4, A) \subseteq t(M, M, I, f^8e, Af^4)$. We have $f^8e =$ $(f^4e)f^4 + f^4D(f^4, e)$. As above,
$t(M, M, I, (f^4D(f^4, e), Af^4) = 0$ and $t(M, M, I, (f^4e)f^4, Af^4) =$
$t(M, M, If^4, f^4e, Af^4)$. Now it remains to repeat the arguments from (1).

(3) It is easy to check that

$$(f^{24}e)e = f^{24}e^2 + 2(ef^{12})^2 - (f^{12}U(e))f^{12} - (e^2f^{12})f^{12}.$$

Hence (3) follows from (2). Lemma is proved.

□

Let T be the ideal of the free Jordan algebra FJ generated by the set $f^{24}(FJ)$. The identity (D) (see the Introduction) implies that an arbitrary multiplication operator of FJ is a linear combination of operators of the type

$$D_1 \cdots D_r,\ D_1 \cdots D_r R(e)R(e),$$

where $e \in FJ$, D_i are inner derivations. We have already mentioned above that $f^{24}(FJ)D_1 \cdots D_r$ lies in the linear span of $f^{24}(FJ)$. Hence, T is spanned by elements $f^{24}(a,b,c), f^{24}(a,b,c)e, \quad (f^{24}(a,b,c)e)e$, where a, b, c, e are arbitrary elements from FJ. Now Lemma 1.1.25 immediatly implies

LEMMA 1.32. $t(M, M, I, T(A), A) = 0$

Remark. In fact, we have proved that for an arbitrary Jordan algebra A, arbitrary A-bimodules M and I and an arbitrary multilinear map $h : M \times M \times I \times A \times A \longrightarrow W$ into a vector space W if h satisfies the conditions (1),(2) and (3) of Lemma 1.29 then $h(M, M, I, T(A), A) = 0$.

LEMMA 1.33. *Suppose that the algebra $\bar{A} = A/I$ is isomorphic to a loop algebra $\mathcal{L}(G)$, where $\mathcal{G}$ is a simple graded finite dimensional Jordan algebra. If $I \neq (0)$ then $\mathcal{G}$ is a Jordan algebra of a bilinear form.*

PROOF. If $\mathcal{G}$ is not an algebra of a bilinear form, then $T(\mathcal{G}) \neq (0)$ (see [Z3]) Hence, $T(\mathcal{L}(G)) \neq (0)$. Since $T(\mathcal{L}(G))$ is a nonzero graded ideal of $\mathcal{L}(G)$ it follows that $T(\mathcal{L}(G)) = \mathcal{L}(G)$ or, equivalently, $A = T(A) + I$. By Lemma 1.32 it implies

$$t(M, M, I, A, A) = t([(MI)A, M]A) = (0).$$

The ideal $I + [(MI)A, M]A$ of A is $\mathcal{D}$-invariant and $t(I + [(MI)A, M]A) = 0$. Hence the zero component of this ideal is nilpotent and therefore
$[(MI)A, M] \subseteq I$.

The last inclusion implies that $id_J(I) \cap A = I$, the contradiction. Lemma is proved. □

LEMMA 1.34. *Let c be an element of A such that its image $\bar{c} \in \bar{A}$ lies in the center of $\bar{A}$. Then $[(MI)c, M] \subseteq I$.*

PROOF. If $\bar{A}$ is a simple finite dimensional algebra or a Jordan algebra of a bilinear form, then $c \in F1 + I$, in which case the assertion is clear. Suppose

that $\bar{A}$ is a positively graded algebra. Choose arbitrary homogeneous elements $x, y \in M$; $u \in I$, $a \in A$. We will show that $t([(xu)c, y]a) = 0$. Indeed, suppose the assertion is not true. Then $t([(xu)c, y]a) = -t([(xu)a, y]c) \neq 0$.

Since the algebra $\bar{A}$ is positively graded, we have $deg(c) \geq 0$. If the degree of c is positive, then the degree of $[(xu)a, y]$ is negative. Hence the element $[(xu)a, y]$ is nilpotent modulo I. Hence, the element $[(xu)a, y]c$ is nilpotent modulo I as well. Nilpotent elements in $\bar{A}_0$ have zero trace. If the degree of c is zero, then $c \in F1 + I$.

This case has already been considered. Thus, $t([(xu)c, y]a) = 0$.

Let C be the preimage of the center of $\bar{A}$ under the homomorphism
$A \to \bar{A}$. Clearly the subspace $[(MI)C, M]$ is $\mathcal{D}$-invariant. By the identity (D) (see the Introduction) the ideal generated by $[(MI)C, M]$ in A is spanned by elements $[(xu)c, y]R(a)R(a)$, where $x, y \in M$; $u \in I$; $c \in C$; $a \in A$. Applying the Jordan identity to $R(y)R(a)R(a)$ we see that

$$[(xu)c, y]R(a)R(a) \in [(MI)C, M]A + [(MI)A, M]$$

Hence, $t(id_A([(MI)C, M])) = (0)$. In view of maximality of I, we have $[(MI)C, M] \subseteq I$.

Now let $\bar{A}$ be isomorphic to a loop algebra $\mathcal{L}(G)$, where $\mathcal{G}$ is a Jordan algebra of a bilinear form $<,>$ in a vector space V, $\mathcal{G} = F.1 + V$. The Z/nZ-gradation on $\mathcal{G}$ is induced by a Z/nZ-gradation on V, $V = \sum_{i=0}^{n-1} V_i$, $\mathcal{G}_0 = F.1 + V_0$, $\mathcal{G}_i = V_i$, $1 \leq i \leq n-1$.

If $V = (0)$ then $\mathcal{L}(G) \simeq F[t^{-n}, t^n]$. If $\dim V = 1$, n is even and $V = V_{n/2}$ then $\mathcal{L}(G) \simeq F[t^{-n/2}, t^{n/2}]$. In all other cases $\mathcal{L}(G)$ contains two orthogonal (not necessarily homogeneous) idempotents $\bar{e}^1$ and $\bar{e}^2$, $\bar{e}^1 + \bar{e}^2 = 1$. Indeed, if $V_0 \neq 0$ then the assertion is well known (see [J3]). Suppose that $V_i \neq 0$, $1 \leq i \leq n-1$, $i \neq \frac{n}{2}$ and the subspaces V_i, V_{n-i} are dual with respect to $<,>$. Choose elements $v_i \in V_i$, $v_{n-i} \in V_{n-i}$ such that $< v_i, v_{n-i} >= 1/2$. Then $\tau = v_i t^i + v_{n-i} t^{-i}$ is an involution in $\mathcal{L}(G)$, $\tau^2 = 1$. Let $\bar{e}^{(1)} = 1/2(1 + \tau)$, $\bar{e}^{(2)} = 1/2(1 - \tau)$.

There remains the case when n is even and $V = V_{n/2}$. If $\dim V \geq 2$ then there exist elements v', $v'' \in V$ such that $< v', v' >=< v'', v'' >= 0$,
$< v', v'' >= 1/2$. Again $\tau = v't^{n/2} + v''t^{-n/2}$ is an involution. Our aim now is to prove that $t([(MI)c, M]A) = (0)$. Since the form $t(ab)$ is nondegenerate on $\mathcal{L}(G)$ this is enough to conclude that $[(MI)c, M] \subseteq I$.

If $\mathcal{L}(G) = F[t^{-n}, t^n]$ or $\mathcal{L}(G) = F[t^{-n/2}, t^{n/2}]$ then we have to prove that $t(M, M, I, c^i, c^j) = 0$; where $c = t^n$ or $c = t^{n/2}$. If i and j are of the same sign then the assertion is clear. The other case boils down to $t(M, M, I, c^{-1}, c)$. Since c is invertible we have $M = MU(c^2)$,
$t(Mc^2, M, I, c^{-1}, c) \subseteq t(M, M, I, c^2c^{-1}, c) + t(Mc^{-1}, M, I, c^2, c) = (0)$.
And similarly $t(Mc^4, M, I, c^{-1}, c) = 0$.

Now suppose that $\mathcal{L}(G)$ contains two orthogonal idempotents $\bar{e}^{(1)}$ and $\bar{e}^{(2)}$, $\bar{e}^{(1)} + \bar{e}^{(2)} = 1$. Let $e^{(1)}, e^{(2)} \in A$ be orthogonal idempotents such that $e^{(i)}$ is a preimage of $\bar{e}^{(i)}$ and $1 = e^{(1)} + e^{(2)}$.

The system of subspaces $\sum_{i \geq k} J_i$ taken for a basis of neighborhoods of zero definitionines a topology on J. Let $\hat{J} = \hat{A} + \hat{M}$ be the completion of J with respect to this topology, let $\hat{I}$ be the closure of the ideal I in $\hat{A}$. Clearly, $\hat{A}/\hat{I} \simeq \widehat{\mathcal{L}(G)}$ is the Jordan algebra of a bilinear form in a vector space over the center $\hat{Z} = F[[t^{-n}, t^n]]$ which is the field of Laurent series. For a proper idempotent in a Jordan algebra of a bilinear form, we have $\mathcal{L}(\hat{G})U(\bar{e}^{(1)}) = \hat{Z}\bar{e}^{(1)}$.

Let a homogeneous element $q \in A$ be a preimage of t^n, $q \in A_n$. The element q is invertible in A and $\hat{A}_{11} = \{e^{(1)}, \hat{A}, e^{(1)}\} \subseteq F[[q^{-1}, q]]e^{(1)} + \hat{I}$.

Let us show that $t(M, M, I, \hat{A}_{11}, \hat{A}_{11}) = (0)$. In view of the inclusion above it is sufficient to prove that $t(M, M, I, q^i e^{(1)}, q^j e^{(1)}) = (0)$. If $i > 0$, $j > 0$ then $q^i \cdot e^{(1)} = (q \cdot e^{(1)})^i \, mod \, I$, $q^j e^{(1)} = (qe^{(1)})^j \, mod \, I$, which implies the result.

For arbitrary i, j choose $k \geq 1$ such that $k+i > 0$, $k+j > 0$. Since the element q is invertible it follows that $M = MU(q^k) \subseteq Mq^k + Mq^{2k}$.

Let $x, y \in M$, $u \in I$. We have $t(xq^k, y, u, q^i e^{(1)}, q^j e^{(1)}) =$
$t(x, y, u, q^k(q^i e^{(1)}), q^j e^{(1)}) - t(x \cdot (q^i \cdot e^{(1)}), y, u, q^k, q^j \cdot e^{(1)})$.

Since $q^k(q^i e^{(1)}) = q^{k+i} e^{(1)} \, mod \, I$ and $t(\cdots, \cdots, \cdots, q^k, q^j e^{(1)}) =$
$t(\cdots, \cdots, \cdots, q^k e^{(1)}, q^j e^{(1)})$ without loss of generality we can assume that $i > 0$. Similarly, we can assume that $j > 0$, which finishes the proof of $t(M, M, I, \hat{A}_{11}, \hat{A}_{11}) = (0)$.

Choose strongly Peirce homogeneous (but not necessarily homogeneous) elements $x, y \in M$, $u \in I$, $a \in A$. Our aim is to prove that $t(x, y, u, c, a) = 0$. Suppose the contrary.

First, we will show that $a \in A_{12} = \{e^{(1)}, A, e^{(2)}\}$. Indeed, suppose that $a \in A_{11}$. We have $c = c_1 + c_2 + c_{12}$, where $c_1 = cU(e^{(1)})$, $c_2 = cU(e^{(2)})$ and $c_{12} = cU(e^{(1)}, e^{(2)}) \in I$. Then $t(x, y, u, a, c) = t(x, y, u, a, c_1) \in t(M, M, I, A_{11}, A_{11}) = (0)$. Similarly, a can not lie in A_{22}. Hence, $a \in A_{12}$.

We have $c = c_1 + c_2 \mod I$. Now

$$t(x,y,u,a,c) = t(x,y,u,a,c_1) + t(x,y,u,a,c_2);$$

$$t(x,y,u,a,c_1) = 2t(x,y,u,ae^{(2)},c_1) = -2t(x,y,u,ac_1,e^{(2)}).$$

By Lemma 1.28 we have $t(M,M,I,A,1) = (0)$. Hence, $t(x,y,u,ac_1,e^{(2)}) = -t(x,y,u,ac_1,e^{(1)})$. Moreover, $a(c_1 - c_2) \in I$, which implies $t(x,y,u,ac_1,e^{(1)}) = t(x,y,u,ac_2,e^{(1)})$.

As a result we get $t(x,y,u,a,c_1) = -t(x,y,u,a,c_2)$ and therefore $t(x,y,u,a,c) = (0)$. Lemma is proved. □

The following Lemma is an analogue of Lemma 1.33 for the case when $\bar{A}$ is a positively graded algebra.

LEMMA 1.35. *If $I \neq (0)$ and $\bar{A}$ is a positively graded algebra such that $\sum_{i\geq s}^{\infty} \mathcal{L}(G)_i \subseteq \bar{A} \subseteq \mathcal{L}(G)$ for some s, then $\mathcal{G}$ is a Jordan algebra of a bilinear form.*

PROOF. Suppose that $\mathcal{G}$ is not a Jordan algebra of a bilinear form. Then $T(\mathcal{G}) \neq (0)$ (see [Z3]). In this case there exists $s' \geq 1$ such that $\sum_{i\geq s'} \mathcal{L}(G)_i \subseteq T(\sum_{i\geq s} \mathcal{L}(G)_i)$.

Let us show that $[(MI)T(A), M] \subseteq I$. Indeed, from Lemma 1.32 it follows that $t([(MI)T(A),M]A) = (0)$. The subspace $I + [(MI)T(A),M]A$ is a graded ideal of A which is $\mathcal{D}$-invariant. This implies the claim.

Hence, for $i \geq s'$ we have $[(MI)A_i, M] \subseteq I$.

Let $\bar{A} = \sum_{i\geq -k}^{\infty} \bar{A}_i$. Suppose that for a homogeneous element $a \in A_i$ and an element $y \in M$ we have $[(MI)a, y] \not\subseteq I$. Then $-k \leq i < s'$.

There exists an element $b \in A_j$, $j \geq s' + k$ such that $[(MI)a, y] \cdot b \not\subseteq I$. By the Jordan identity we have $(MI)R(a)R(y)R(b) \subseteq (MI)(-R(b)R(y)R(a) - R(y(ab)) + R(ya)R(b) + R(yb)R(a) + R(ab)R(y)) \subseteq I$, the contradiction. Lemma is proved. □

Define the mapping $h_k : M \times M \times \underbrace{I \times \cdots \times I}_{k} \times \underbrace{A \times \cdots \times A}_{k} \longrightarrow \bar{A}$ as follows.
For $x, y \in M$; $u_1, \ldots, u_k \in I$; $a_1, \ldots, a_k \in A$,we let

$$h_k(x,y,u_1,\ldots,u_k,a_1,\ldots,a_k) = [x\prod_{i=1}^{k} R(u_i)R(a_i), y] + I/I.$$

LEMMA 1.36. *(1) $h_k(x,y,u_1,\ldots,u_k,a_{\sigma(1)},\ldots,a_{\sigma(k)}) = (-1)^{\sigma} h_k(x,y,u_1,\ldots,u_k,a_1,\ldots,a_k)$ for an arbitrary permutation σ.*

(2) $h_k(x,y,u_1,\ldots,u_k,a_1^2,a_2,\ldots,a_k) = 2h_k(xa_1,y,u_1,\ldots,u_k,a_1,\ldots,a_k) = 2h_k(x,ya_1,u_1,\ldots,u_k,a_1,\ldots,a_k) = 2h_k(x,y,u_1,\ldots,u_ja_1,\ldots,u_k,a_1,\ldots,a_k) = 2h_k(x,y,u_1,\ldots,u_k,a_1,\ldots,a_ja_1,\ldots,a_k)$ for any $1 \leq j \leq k$.

(3) If some of the elements a_i is central modulo I, then $h_k(x, y, u_1, \ldots, u_k, a_1, \ldots, a_k) = 0$.

PROOF. We have already mentioned above that

$$R(I)R(I)R(A) \subseteq R < A > R(I).$$

From the identity (D) (see the Introduction) and again the Jordan identity it follows that

$$R(I)R(I)R < A > \subseteq R < A > R(I).$$

(1) If $a_i = a_{i+1} = a$, then $R(u_i)R(a)R(u_{i+1})R(a) = \frac{1}{2}R(u_i)(-R(u_{i+1}a^2) + 2R(u_{i+1}a)R(a)) \in R < A > R(I)$. Hence $\prod_{i=1}^k R(u_i)R(a_i) \in \cdots R(I)R(I)R < A > \subseteq R < A > R(I)$ and therefore $[x \prod_{i=1}^k R(u_i)R(a_i), y] \in [MI, y] \subseteq I$.

(2) By the Jordan identity $R(u_1)R(a_1^2) = -2R(a_1)R(u_1a_1) + 2R(u_1a_1)R(a_1) + R(a_1^2)R(u_1)$, so $R(u_1)R(a_1^2) = 2R(u_1a_1)R(a_1) \mod R < A > R(I)$.

This implies

$$h_k(x, y, u_1, \ldots, u_k, a_1^2, a_2, \ldots, a_k) = 2h_k(x, y, u_1a_1, \ldots, u_k, a_1, \ldots, a_k).$$

We have
$xR(u_1)R(a_1^2) = u_1R(x)R(a_1^2) = u_1(-2R(a_1)R(xa_1) + 2R(xa_1)R(a_1) + R(a_1)^2R(x)) = -2xR(a_1)R(u_1a_1) + xR(u_1a_1^2) + 2(xa_1)R(u_1)R(a_1) = 2(xa_1)R(u_1)R(a_1) \mod MI$.
This implies that

$$h_k(x, y, u_1, \ldots, u_k, a_1^2, a_2, \ldots, a_k) = 2h_k(xa_1, y, u_1, \ldots, u_k, a_1, \ldots, a_k).$$

We have also

$$R(u_k)R(a_k^2)R(y) = -2R(u_k)R(ya_k)R(a_k) + 2R(u_k)R(a_k)R(ya_k) + R(u_k)R(y)R(a_k^2).$$

This implies that

$$h_k(x, y, u_1, \ldots, u_k, a_1, \ldots, a_k^2) = 2h_k(x, ya_k, u_1, \ldots, u_k, a_1, \ldots, a_k).$$

And, finally it remains to prove that $h_k(x, y, u_1, \ldots, u_k, a_1, a_1^2, \ldots, a_k) = 0$. But, $h_k(x, y, u_1, \ldots, u_k, a_1, a_1^2, \ldots, a_k) = 2h_k(xa_1, y, u_1, \ldots, u_k, a_1, a_1, \ldots, a_k) = 0$. (3) If one of the elements a_i is central modulo I then by (1) we can assume that a_k is central modulo I. Then the assertion follows from Lemma 1.34. Lemma is proved. □

LEMMA 1.37. *Suppose that the algebra $\bar{A}$ is finitely generated. Then there exists $k \geq 1$ such that $h_k = 0$.*

PROOF. Let elements $b_1, \ldots, b_d \in A$ generate A modulo I. Using Lemma 1.36(2) for any $k \geq 1$ we have $h_k(M, M, I, \ldots, I, A, \ldots, A) = \sum h_k(M, M, I, \ldots, I, b_{i_1}, \ldots, b_{i_k}), 1 \leq i_1, \ldots, i_k \leq d)$. This and Lemma 1.36(1) imply that $h_{d+1} = 0$. Lemma is proved. □

LEMMA 1.38. *Let $\bar{A}$ be a positively graded algebra, let $x, y \in M$; $u_1, \dots, u_k \in I$; $a_1, \dots, a_k \in A$ be elements such that $h_k(x, y, u_1, \dots, u_k, a_1, \dots, a_k) \neq 0$. Let $r \geq 1$ be an arbitrary integer. Then there exists an element $a' \in \sum_{i=r}^{\infty} A_i$ such that $h_k(x, y, u_1, \dots, u_k, a_1, \dots, a_{k-1}, a') \neq 0$.*

PROOF. Since $[x \prod_{i=1}^{k} R(u_i)R(a_i), y] \notin I$, there exists an element $b \in A$ of arbitrarily big degree such that $[x \prod_{i=1}^{k} R(u_i)R(a_i), y] \cdot b \notin I$. From $R(a_k)R(y)R(b) = -R(b)R(y)R(a_k) - R(y(ba_k)) + R(ya_k)R(b) + R(a_k b)R(y) + R(yb)R(a_k)$ it follows that $h_k(x, y, u_1, \dots, u_k, a_1, \dots, a_{k-1}, b) \neq 0$ or $h_k(x, y, u_1, \dots, u_k, a_1, \dots, a_{k-1}, a_k b) \neq 0$. Lemma is proved. $\square$

Let K denote the ideal of A generated by $[(MI)A, (MI)A]$.

LEMMA 1.39. *$KM \subseteq (MI)A$.*

PROOF. The subspace $M' = (MI)A$ is an A-subbimodule of M. This implies that $K = [M', M']A$.

Indeed, for arbitrary elements $m', m'' \in M'$; $a, b \in A$ we have $m'R(m'')R(a)R(b) = m'(-R(b)R(a)R(m'') - R((m''b)a) + R(m'')R(ab) + R(a)R(m''b) + R(b)R(m''a)) \in [M', M']A$.

The subbimodule M' is $\mathcal{D}$-invariant. Hence for an arbitrary element $m \in M$ we have $[m', m'']m = m'D(m'', m) - [m', m] \cdot m'' \in M'$.

And $[m', m'']R(a)R(m) = [m', m'']D(a, m) + ([m', m'']m)a = m'R(m''D(a, m)) - m''R(m'D(a, m)) + ([m', m'']m)a \in M'$.
Lemma is proved. $\square$

The ideal K of A is $\mathcal{D}$–invariant. Suppose that K does not lie in I. Then the zero component K_0 of K is not nilpotent and thus contains an idempotent e. By Lemma 1.39 $Me \subseteq M'$. Let $J = J_{11} + J_{10} + J_{00}$ be the Peirce decomposition of J with respect to the idempotent e, $J_{11} = JU(e)$, $J_{10} = JR(1-e)R(e)$, $J_{00} = JU(1-e)$.

LEMMA 1.40. *For any $k \geq 2$ we have $h_k(M_{00}, M_{00}, I, \dots, I, A, \dots, A) = (0)$.*

PROOF. Choose elements $x, y \in M_{00}$; $u_1, \dots, u_k \in I$; $a_1, \dots, a_k \in A$. Without loss of generality we assume that all elements a_i are homogeneous and strongly Peirce homogeneous. If $a_i \in A_{11}$ then $h_k(x_{00}, \dots, a_i e, \dots) = h(x_{00}a_i, \dots) + h(x_{00}e, \dots) = 0$. If $a_i \in A_{10}$ then $a_i = 2a_i e$ and $h_k(x_{00}, y_{00}, \dots, a_i e, \dots)$

$= h(x_{00}a_i, y_{00}, \dots, e, \dots) = 2h_k(x_{00}a_i, y_{00}e, \dots, e, \dots) = 0.$

Hence all elements a_i can be assumed to lie in A_{00}.

Let $\bar{A} \simeq \mathcal{L}(G)$, where $\mathcal{G}$ is a $\mathbb{Z}/n\mathbb{Z}$-graded Jordan algebra of a bilinear form. Then $\bar{A}_{00} = \bar{A}U(1-\bar{e}) = F[t^{-n}, t^n](1-\bar{e})$. Hence, $\bar{a}_2 = \alpha \bar{a}_1 t^{ni}$, $\alpha \in F$. Let c be a preimage of αt^{ni}.
We have $h_k(x, y, u_1, \dots, u_k, a_1, \dots a_k) = h_k(x, y, u_1, \dots, u_k, a_1, a_1 c, \dots, a_k) = \frac{1}{2} h_k(x, y, u_1, \dots, u_k, a_1^2, c, a_3, \dots a_k) = 0$ by Lemma 1.34.

If $\bar{A}$ is a simple finite or infinite dimensional algebra of a bilinear form, then dim $\bar{A}_{00} = 1$ and so a_1 and a_2 are linearly dependent modulo I.

Now let $\bar{A}$ be a positively graded algebra, $\sum_{i=s}^{\infty} \mathcal{L}(G)_i \subseteq \bar{A} \subseteq \mathcal{L}(G)$, $\mathcal{G}$ is a $\mathbb{Z}/n\mathbb{Z}$-graded Jordan algebra of a bilinear form. In view of Lemma 1.38 we may assume that $deg(a_k) - deg(a_1) \geq s$. Since both $\bar{a}_1$ and $\bar{a}_k$ lie in $F[t^{-n}, t^n](1 - \bar{e})$ it follows that $\bar{a}_k = \alpha \bar{a}_1 t^{ni}$, $ni \geq s$, $\alpha \in F$, .

We have $h_k(x, y, u_1, \dots, u_k, a_1, \dots a_k) = h_k(x, y, u_1, \dots, u_k, a_1, \dots a_1 c) = \frac{1}{2} h_k(x, y, u_1, \dots, u_k, a_1^2, a_2, \dots, a_{k-1}, c) = 0$. Lemma is proved. □

LEMMA 1.41. *Suppose that $\bar{A}$ is a finitely generated algebra. Then* $[(MI)A, (MI)A] \subseteq I$.

PROOF. If $[M', M'] \not\subseteq I$, then $K \not\subseteq I$, which was our starting assumption for the existence of the idempotent e and for Lemma 1.40.

It is easy to see that $[(MI)A, (MI)A] = [MD(I, A), MD(I, A)] \bmod I = [MD(I, A)D(I, A), M] \bmod I$.

Suppose that the Lemma is wrong. Then by Lemma 1.1.31 there exists $k \geq 2$ such that $[M(R(I)R(A))^k, M] \not\subseteq I$, but $[M(R(I)R(A))^{k+1}, M] \subseteq I$.

Choose Peirce homogeneous (with respect to e) elements $x, y \in M$ such that $[x(R(I)R(A))^k, y] \not\subseteq I$. Then $x \notin MR(I)R(A)$, which implies $x \in M_{00}$. Indeed, $M_{11} + M_{10} = M \cdot e \subseteq M'$ by Lemma 1.39.

For arbitrary elements $u_1, \dots, u_k \in I$; $a_1, \dots, a_k \in A$ we have $[x \prod_{i=1}^k R(u_i)R(a_i), y] = [x \prod_{i=1}^k D(u_i, a_i), y] \; mod\, I = (-1)^{k+1}[yD(u_k, a_k) \cdots D(u_1, a_1), x] \; mod\, I$, which implies that $y \in M_{00}$ as well. This contradicts Lemma 1.40. Lemma is proved. □

Now we are ready to summarize what we did in this chapter

PROPOSITION 1.42. Let $J = A + M$ be a graded simple Jordan superalgebra with dimensions $dim(J_i)$ uniformely bounded. Let I be the maximal $\mathcal{D}$-invariant graded ideal of A such that I_0 is nilpotent. Then one of the following statements holds:

(1) $I = M(A) \neq (0)$, $A/M(A) \simeq \mathcal{L}(G)$, a loop algebra of a simple finite dimensional Jordan algebra of a bilinear form, or

(2) $I = M(A) \neq 0$, A/I is an infinite dimensional Jordan algebra of a bilinear form, or

(3) $I = M(A) \neq 0$, A/I is a finite dimensional simple Jordan algebra of a bilinear form, or

(4) A/I is a one-sided graded Jordan algebra, $\sum_{i=s}^{\infty} \mathcal{L}(G)_i \subseteq A/I \subseteq \mathcal{L}(G)$,where $\mathcal{G}$ is a Jordan algebra of a bilinear form, (the case $I = (0)$ is included), or

(5) $I = (0)$, $A = A^{(1)} \oplus A^{(2)}$, where $A^{(i)}$ are prime nondegenerate algebras, or

(6) $A \simeq \mathcal{L}(G)$ is a loop algebra of a simple finite dimensional Jordan algebra, or

(7) A is a finite dimensional simple algebra, or

(8) A is an infinite dimensional simple Jordan algebra of a bilinear form.

PROPOSITION 1.43. (1) $[MI, M] \subseteq I$,

(2) If an element $c \in A$ is central modulo I, then $[(MI)c, M] \subseteq I$,

(3) If $\bar{A}$ is a finitely generated algebra then $[(MI)A, (MI)A] \subseteq I$ (or equivalently, $[(((MI)A)I)A, M] \subseteq I$).

CHAPTER 2

Cartan type

2.1. Results

DEFINITION 2.1. A graded simple Lie (super)-algebra L is said to be of **Cartan type** if it contains a proper graded subalgebra L' of finite codimension.

A simple $\mathbb{Z}$-graded Lie superalgebra $L = L_{\bar{0}} + L_{\bar{1}}$ of Cartan type can not contain the Virasoro algebra Vir in its even part. Indeed, let $Vir \subseteq L_{\bar{0}}$. Let $L' = L'_{\bar{0}} + L'_{\bar{1}}$ be a proper subsuperalgebra of L of finite codimension. Since the Virasoro algebra does not have proper subalgebras of finite codimension we have $Vir \subseteq L'_{\bar{0}}$. Then the quotient space L/L' becomes a module over Vir. Since Vir does not have finite dimensional modules with nonzero action we have $[L, Vir] \subseteq L'$. This easily implies that the ideal generated by Vir in L lies in L', which contradicts the simplicity of L.

DEFINITION 2.2. A graded simple Jordan superalgebra J is of **Cartan type** if it contains a proper subalgebra B such that B is of finite codimension and $B^- + [B^-, J^+] + [J^-, B^+] + B^+$ is a subalgebra of finite codimension in the Tits-Kantor-Koecher Lie (super)-algebra $K(J)$.

As in the previous chapter we study a $\mathbb{Z}$-graded simple Jordan superalgebra $J = A + M$ having the dimensions of all components J_i uniformely bounded.

The main results of this chapter are:

PROPOSITION 2.3. If A/I is a one-sided graded algebra, then J is of Cartan type (regardless of whether $I \neq (0)$ or $I = (0)$.)

PROPOSITION 2.4. If $I = (0)$, $A = A^{(1)} \oplus A^{(2)}$, $A^{(1)}$ is finite dimensional and $A^{(2)}$ is a one sided graded algebra, then J is of Cartan type.

PROPOSITION 2.5. If $I = (0)$, $A = A^{(1)} \oplus A^{(2)}$ and both algebras $A^{(1)}$, $A^{(2)}$ are positively (resp. negatively) graded, then J is of Cartan type.

2.2. A/I one-sided graded

Now we will assume that $A/I = \sum_{i \geq -k}^{\infty} \bar{A}_i$ is positively graded (the case of a negative gradation is similar).

Let $\bar{C} = \sum_{i\geq 0} \bar{C}_i$ be the center of the algebra $\bar{A}$. If for an arbitrary element $\bar{c} \in \bar{C}_i$, $i > 0$, an arbitrary derivation $D(x,y)$, $x, y \in M$, the element $\bar{c}D(x,y)$ has positive degree or is equal to 0, then the ideal of $\bar{A}$ generated by $\bar{C}_+ = \sum_{i>o} \bar{C}_i$ is $\mathcal{D}$-invariant. The (k+1)-th power of this ideal will have zero intersection with $\bar{A}_0$, which contradicts maximality of I.

Hence, there exists a homogeneous element $c \in A$ and homogeneous elements $x, y \in M$ such that the element $\bar{c}$ is central in $\bar{A}$ and $cD(x,y) = 1 \bmod I_0$. Hence,

(2.1)
$$M = MR(cD(x,y)) \subseteq MD(x,y) + MD(x,y)R(c) \subseteq xA + yA + (xA)c + (yA)c$$

From (2.1) and from $A = \sum_{i\geq -k}^{\infty} A_i + I$ it follows that there exists a number $q \geq 1$ such that

$$M = xI + yI + (xI)c + (yI)c + \sum_{i\geq -q} M_i \tag{2.2}$$

Let $B = B_{\bar{0}} + B_{\bar{1}}$, $B_{\bar{0}} = I + \sum_{i\geq 1} A_i$, $B_{\bar{1}} = (MI)A + \sum_{i\geq q+1} M_i$.

LEMMA 2.6. *B is a subalgebra of J.*

PROOF. In view of Lemma 1.41 the only inclusion that needs verification is $[(MI)A, \sum_{i\geq q+1} M_i] \subseteq B_{\bar{0}}$.

But (2.2) and Lemma 1.34 imply that $[M, \sum_{i\geq q+1} M_i] \subseteq B_{\bar{0}}$, which is an even stronger inclusion. Lemma is proved. □

LEMMA 2.7. *B is of finite codimension in J.*

PROOF. From $A = \sum_{i\geq -k}^{\infty} A_i + I$ it follows that $B_{\bar{0}}$ is of finite codimension in A. From (2.2) it follows that $B_{\bar{1}}$ is of finite codimension in M. Lemma is proved. □

Let $B' = B'_{\bar{0}} + B'_{\bar{1}}$, where

$$B'_{\bar{0}} = I + \sum_{i\geq 2max(q,k)+1} A_i,$$

$$B'_{\bar{1}} = MI + \sum_{i\geq 2max(q,k)+1} M_i + (MI)c.$$

It is easy to see that $B'J \subseteq B$ and B' is still of finite codimension in J.

LEMMA 2.8. *There exists a number $s \geq 1$ such that*

$$[(MI + (MI)c) \sum_{i\geq s} A_i, M] \subseteq B_{\bar{0}}.$$

PROOF. For $s \geq q+1$ we have $(MI + (MI)c) \sum_{i\geq s} A_i \subseteq B_{\bar{1}}$. Since $M = B_{\bar{1}} + \sum_{j=-q}^{q} M_j$ and B is a subsuperalgebra, it is sufficient to check that $[(MI + (MI)c) \sum_{i\geq s} A_i, \sum_{j=-q}^{q} M_j] \subseteq B_{\bar{0}}$.

As we have already noticed in Chapter I (see the proof of Lemma 1.13), there exists $s \geq 1$ such that $\sum_{i\geq s} A_i \subseteq (\sum_{i\geq 2q+1} A_i)^2 + I$. Choose arbitrary elements

$a \in \sum_{i\geq 2q+1} A_i$, $x \in \sum_{j=-q}^{q} M_j$. We claim that $[((MI)c + MI)a^2, x] \in B_{\bar{0}}$.

Indeed, $R(a^2)R(x) = 2[R(a), R(ax)] + R(x)R(a^2)$.

Both elements a and ax lie in B and $[MI + (MI)c, x]a^2 \in I$ by Lemma 1.34. Lemma is proved. □

Now let $B'' = B''_{\bar{0}} + B''_{\bar{1}}$,where $B'_{\bar{0}} = I + \sum_{i\geq max(s,2max(q,k)+1)} A_i$, $B''_{\bar{1}} = B'_{\bar{1}}$.

Lemma 2.9. $(B''B')J \subseteq B$.

Proof. The only thing that does not allow us to say that B' is a subsuperalgebra is that it is not clear if $(MI + (MI)c)\sum_{i\geq 2max(q,k)+1} A_i$ lies in $B'_{\bar{1}}$.

The products of any other two components of B' lie in B'. Thus we have to check only that $((MI + (MI)c)\sum_{i\geq max(s,2max(q,k)+1)} A_i)J \subseteq B$. But this follows from Lemma 2.8. Lemma is proved. □

Remark. B'' is of finite codimension in J.

The quotient $V = J/B$ is a finite dimensional bimodule over B. Let $B''' = \{b \in B'' | (bB)V = (0)\}$.

Lemma 2.10. *B''' is of finite codimension in J.*

Proof. By Lemma 2.9, for an arbitrary element $b \in B''$ we have $(bB')V = (0)$.

Let $\{b_1, \ldots, b_n\}$ be a basis of B modulo B' and let $\{v_1, \ldots, v_t\}$ be a basis of V. Then $B''_{ij} = \{b \in B'' | (bb_i)v_j = 0\}$ has finite codimension in B'', because $dim_F V < \infty$. Hence $B''' = \cap_{i,j} B''_{ij}$ has finite codimension in B'' and in J. Lemma is proved. □

Consider the annihilator $\tilde{B} = Ann_B(V) = \{b \in B | bV = (0), BD(b,V) = (0)\} = \{b \in B | bV = (0), (bB)V = (0)\}$.

From [Z1] it follows that $\tilde{B}$ is an ideal of B. Moreover, the inclusions $B''' \subseteq B'$ and $B'J \subseteq B$ imply that $B''' \subseteq \tilde{B}$. Hence the ideal $\tilde{B}$ of B is of finite codimension.

Lemma 2.11. *$K_J(\tilde{B}) = \tilde{B}^- + [\tilde{B}^-, J^+] + [\tilde{B}^+, J^-] + \tilde{B}^+$ is a subsuperalgebra of $K(J)$ of finite codimension.*

Proof. To prove the first part of the statement we have to show that $\{\tilde{B}, \tilde{B}, J\} \subseteq \tilde{B}$ and $\{\tilde{B}, J, \tilde{B}\} \subseteq \tilde{B}$.

We will show that $(\tilde{B}J)\tilde{B} \subseteq \tilde{B}$ and $(\tilde{B}\tilde{B})J \subseteq \tilde{B}$.

The first inclusion is easier: $\tilde{B}J \subseteq B$, $B\tilde{B} \subseteq \tilde{B}$. As for the second inclusion, we know that $(\tilde{B}\tilde{B})J \subseteq B$. It remains to prove that $(\tilde{B}\tilde{B})J$ annihilates J modulo B. In other words, we have to prove that $((\tilde{B}\tilde{B})J)J \subseteq B$ and $(((\tilde{B}\tilde{B})J)B)J \subseteq B$.

From the Jordan identity it follows that
$\tilde{B}R(\tilde{B})R(J)R(J) \subseteq \tilde{B}(R(J)R(J)R(\tilde{B}) + R((\tilde{B}J)J) + R(\tilde{B}J)R(J) + R(J)R(\tilde{B}))$.

Examining the summands on the right hand side we get:

$\tilde{B}(R(J)R(J)R(\tilde{B}) + R((\tilde{B}J)J) + R(J)R(\tilde{B})) \subseteq J\tilde{B} \subseteq B$ and $\tilde{B}R(\tilde{B}J)R(J) \subseteq \tilde{B}R(B)R(J) \subseteq \tilde{B}J \subseteq B$.

Let's prove the second inclusion. We have,

$\tilde{B}R(\tilde{B})R(J)R(B)R(J) \subseteq \tilde{B}(R(B)R(J)R(\tilde{B}) + R((\tilde{B}B)J) + R(\tilde{B}J)R(B) + R(\tilde{B}B)R(J) + R(BJ)R(\tilde{B}))R(J)$.

Again we have to analize summands on the right hand side.

$\tilde{B}R(B)R(J)R(\tilde{B})R(J) \subseteq \tilde{B}R(J)R(\tilde{B})R(J) \subseteq BR(\tilde{B})R(J) \subseteq \tilde{B}J \subseteq B$,

$\tilde{B}R((\tilde{B}B)J)R(J) \subseteq \tilde{B}R(B)R(J) \subseteq \tilde{B}J \subseteq B$,

$\tilde{B}R(\tilde{B}J)R(B)R(J) \subseteq \tilde{B}R(B)R(J) \subseteq \tilde{B}J \subseteq B$,

$\tilde{B}R(\tilde{B}B)R(J)R(J) \subseteq (\tilde{B}\tilde{B})R(J)R(J) \subseteq B$, (see above);

$\tilde{B}R(BJ)R(\tilde{B})R(J) \subseteq \tilde{B}R(J)R(\tilde{B})R(J) \subseteq (\tilde{B}B)J \subseteq \tilde{B}J \subseteq B$.

This proves that $K_J(\tilde{B})$ is a subsuperalgebra of $K(J)$.

Let $\{w_1, \dots, w_d\}$ be a basis of J modulo $\tilde{B}$. Then the elements w_i^-, w_i^+, $[w_i^-, w_i^+]$ span $K(J)$ modulo $K_J(\tilde{B})$. So $K_J(\tilde{B})$ has finite codimension in $K(J)$. Lemma is proved. □

We have proved Proposition 2.3.

Exactly in the same way as above we can prove that J is of Cartan type when $I = (0)$, $A = A^{(1)} \oplus A^{(2)}$, and

(i) $A^{(1)}$ is finite dimensional and $A^{(2)}$ is one-sided graded, or

(ii) $A^{(1)}$ and $A^{(2)}$ are both one-sided graded with positive (resp. negative) gradation.

In these cases, we will consider a central element of positive degree in $A^{(2)}$ and will use the fact that $M = \{e, M, f\}$, where e is the identity in $A^{(1)}$ and f is the identity in $A^{(2)}$.

CHAPTER 3

Even Part is Direct Sum of two Loop Algebras

3.1. General Results

In this chapter we will take care of the following cases:

1. $A = F[t^{-n}, t^n]$ is the algebra of Laurent polynomials in t^n,

2. $A = F[t^{-n}, t^n] \oplus F[t^{-m}, t^m]$ is a direct sum of two algebras of Laurent polynomials,

3. $A = \mathcal{L}(\mathcal{G}') \oplus \mathcal{L}(\mathcal{G}'')$ is a direct sum of two loop algebras; $\mathcal{G}'$, $\mathcal{G}''$ are finite dimensional simple graded Jordan algebras, graded by a finite cyclic group. But first we will prove a proposition that applies to all three cases and also to chapter 4.

PROPOSITION 3.1. Let A be a loop algebra or a sum of two loop algebras. Let $Z(A)$ be the center of A and suppose that $Z(A)D(M, M) = (0)$. Then J is a loop superalgebra, $J \simeq \mathcal{L}(G)$, where $\mathcal{G}$ is a finite dimensional superalgebra.

Remark that if A is a loop algebra or a direct sum of two loop algebras then $Z(A)$ contains a homogeneous invertible element c of positive degree n (see [J3] about invertibility in Jordan algebras).

As above, by $R < J >$ we denote the multiplication algebra of the subalgebra J.

Now our aim is to prove that $R < J >$ is a finitely generated free right module over the subalgebra $F[U(c^{-1}), U(c)]$.

Let S denote the subalgebra of $R < J >$ generated by $D(A, A)$, $D(M, M)$ and $R(c^i)$, $i \in \mathbb{Z}$.

LEMMA 3.2. *$R < J >$ is a finitely generated right module over S.*

PROOF. For an arbitrary element $x \in A \cup M$ we have

$$R(xU(c^2)) = 2V(c, xU(c))R(c) - V(c^2, x)U(c) \tag{3.1}$$

Let $R << c >>$ denote the subalgebra of $R < J >$ generated by multiplications $R(c^i)$, $i \in \mathbb{Z}$.

Let $y \in J_k$, where $k \geq 4n$. Then $y = xU(c^2)$, $0 \leq deg(x) = deg(y) - 4n$. From (3.1 it follows that $R(y) \in \sum_i R(y_i)R << c >> + R(c)\sum_i R(y'_i)R << c >>$ $+ R(c^2)\sum_i R(y''_i)R << c >>$, where $0 \leq deg(y_i) \leq k - n$, $0 \leq deg(y'_i) \leq k - 2n$,

$0 \le deg(y_i'') \le k - 4n$. The third summand on the right hand side can be ignored because of the equality $R(c^2)R(y''_i)0 - 2R(y''_i c)R(c) + R(y''_i c^2) + 2R(c)R(y''_i)R(c)$.

Let us use induction on $deg(y) = k$ to show that

$$R(y) \in \sum_{k=0}^{4n-1} R(J_i)R << c >> + R(c) \sum_{k=0}^{4n-1} R(J_i)R << c >> .$$

Applying the induction assumption to the elements y_i, y_i' above, we are left only with operators of the type $R(c)^2 R(z)$, where $0 \le deg(z) \le min(k - 4n, 4n - 1)$. We have $[R(c), [R(c), R(z)]] = R(cD(c, z))$. This implies $R(c)^2 R(z) = R(cD(c, z)) - R(z)R(c)^2 + 2R(c)R(z)R(c)$, and it remains to apply the induction assumption to $R(cD(c, z))$. Similarly, if $deg(y) < 0$ thn $R(y) \in \sum_{i=0}^{4n-1} R(J_{-i})R << c >>$ $+R(c)\sum_{i=0}^{4n-1} R(J_{-i})R << c >>$.

Hence $R(J)$ lies in a finitely generated right $R << c >>$-module. Since $R << c >>$ is Noetherian it follows that there exists a finite collection of elements $u_i \in A \cup M$ such that $R(J) \subseteq \sum_i R(u_i)R << c >>$.

Now, $R(J)R(A) \subseteq \sum R(u_i)R << c >> R(u_j)R << c >>$, $u_j \in A$. Since $[R(c^l), [R(c^k), R(u_j)]] = 0$ for $u_j \in A$ we conclude that

$$R(J)R(A) \subseteq \sum R(u_i)R(c^k)R(u_j)R << c >>, \ |k| \le 2.$$

So again $R(J)R(A)$ lies in a finitely generated right $R << c >>$-module.

Similarly, $R(A)R(J) \subseteq \sum R(u_j)R << c >> R(J)$, $u_j \in A$. As above from $[R(c^l), [R(c^k), R(J)]] \subseteq R(J)$ it follows that
$R(A)R(J) \subseteq \sum R(u_j)R(c^k)R(J)R<<c>> \subseteq \sum R(u_j)R(c^k)R(u_i)R<<c>>$,
$|k| \le 2$.

We have shown that $D(M, A) \subseteq R(M)R(A) + R(A)R(M)$ lies in a finitely generated right $R << c >>$-module.

Consider the Lie superalgebra $L = R(J) + D(J, J)$. We proved that there exists a finite collection of elements $v_i \in L$ such that $L \subseteq \sum_i v_i S$. It will be convenient to assume that the set $\{v_i\}_i$ includes the identity operator.

We claim that for an arbitrary $k \ge 1$, $L^k \subseteq \sum v_{i_1} \cdots v_{i_k} S$. Indeed, for $k = 1$ the assertion is true. If $k > 1$ then $L^k \subseteq L^{k-1} \sum_i v_i S$. We have $L^{k-1}v_i \subseteq v_i L^{k-1} + L^{k-1}$. Hence, $L^k \subseteq \sum v_i v_{i_1} \cdots v_{i_{k-1}} S + \sum v_{i_1} \cdots v_{i_{k-1}} S$, by the induction assumption. This proves the claim.

Let A be generated by m elements $a_1, \ldots, a_m$. Then $D(M, A)^{2m+1} \subseteq R(J)^{4m+1}$. Indeed, because of the identity $D(x, ab) = D(xa, b) + D(xb, a)$ it is sufficient to prove that $D(x_1, a_{i_1}) \cdots D(x_{2m+1}, a_{i_{2m+1}}) \in R(J)^{4m+1}$ for arbitrary elements $x_1, \ldots, x_{2m+1} \in M$.

At least three of the elements $a_{i_1}, \ldots, a_{i_{2m+1}}$ are equal. Putting the corresponding superderivations together (modulo shorter products) we get an operator of the type $D(x, a)D(y, a)D(z, a)$, $x, y, z \in M$, inside our product. Now the Jordan identity implies that $D(x, a)D(y, a)D(z, a) \in R(J)^5$.

From the identity (D) (see the Introduction) it follows that an arbitrary multiplication operator $R(x_1) \cdots R(x_r)$, $x_i \in A \cup M$ is a linear combination of the operators of the type

$$R(y_1)\cdots R(y_\mu)D_1\cdots D_l d_1\cdots d_q,$$

where $y_i\in A\cup M$, $0\le\mu\le 2$, $D_i\in D(M,A)$, $d_j\in D(A,A)+D(M,M)$ and $\mu+2(l+q)\le r$.

As we have seen above, we can assume that $l\le 2m$. Hence $R<J>=L^{2m+2}S=\sum v_{i_1}\cdots v_{i_{m+2}}S$. Lemma is proved. □

LEMMA 3.3. *The algebra $R<J>$ is PI.*

PROOF. From $Z(A)D(M,M)=(0)$ it follows that the operators $U(c)$, $U(c^{-1})$ lie in the center of S.We have noticed in the proof of Lemma 3.2 that the vector space J is a finitely generated $F[U(c),U(c^{-1})]$-module. Besides, this module is graded, hence torsion free, hence, J is a free module over the PID $F[U(c),U(c^{-1})]$, say of rank r. Hence, S is embeddable into the matrix algebra $\mathcal{M}_r(F[U(c),U(c^{-1})])$ which implies that S is a free $F[U(c),U(c^{-1})]$-module of rank $\le r^2$. By Lemma 3.2 $R<J>$ is a finitely generated right $F[U(c),U(c^{-1})]$-module and for the same reason as above it is free. If l is the rank of this module, then $R<J>$ is embeddable into the matrix algebra $\mathcal{M}_l(F[U(c),U(c^{-1})])$. By the Amitsur-Levitzky theorem (see [R2]) $R<J>$ is P.I. Lemma is proved. □

REMARK 3.4. Any $\mathbb{Z}$-graded graded prime algebra R is prime.

Indeed, for an ideal I of R let I_{gr} be the linear span of all highest degree homogeneous components of I. Then I_{gr} is a graded ideal of R. If I, I' are (not necessarily) graded nonzero ideals of R such that $I\cdot I'=(0)$ then $I_{gr}\cdot I'_{gr}=(0)$, which contradicts R being graded prime.

LEMMA 3.5. *Let $R=\sum_{i\in\mathbb{Z}}R_i$ be a $\mathbb{Z}$-graded prime associative PI-algebra with infinitely many nonzero homogeneous components R_i. Then R contains a nonzero central element of nonzero degree.*

PROOF. It is easy to see that the center $Z(R)$ is a graded subalgebra of R. Suppose that $Z(R)\subseteq R_0$. It is known (see [R1] ,[M]) that $Z(R)\ne(0)$ and the ring of fractions $(Z(R)-\{0\})^{-1}R$ is a simple finite dimensional algebra over the field of fractions of $Z(R)$.

From $Z(R)\subseteq R_0$ it follows that the algebra $(Z(R)-\{0\})^{-1}R$ is graded and $(Z(R)-\{0\})^{-1}R=\oplus_{i\in\mathbb{Z}}(Z(R)-\{0\})^{-1}R_i$ is a direct sum of vector spaces over the field $(Z(R)-\{0\})^{-1}Z(R)$. Since the left hand side of this equality is finite dimensional it follows that the right hand side contains finitely many summands. Hence only finitely many components R_i are not equal to zero. Lemma is proved. □

Now we are ready to finish the proof of Proposition 3.1. The algebra $R<J>$ has a graded irreducible faithful module J. Hence $R<J>$ is graded prime, hence it is prime. From Lemmas 3.3 and 3.5 it follows that if A is a loop algebra or a sum of two loop algebras and $Z(A)D(M,M)=(0)$, then $R<J>$ contains a central element ω of nonzero degree. The algebra $R<J>$ is $\mathbb{Z}/2\mathbb{Z}$-graded and the center of $R < J >$ is a graded subspace. Hence, without loss of generality we can assume that ω is odd or even. For an arbitrary element $x \in A \cup M$ we have $x\omega = 1R(x)\omega = 1\omega R(x)$. Hence $\omega = R(1\omega)$. Denote $z = 1\omega$. Clearly, $R(z)^2 = R(z^2)$. If $z \in M$ then $R(z)^2 = 0$. Since the algebra $R < J >$ does not contain nonzero nilpotent central elements, it follows that $z \in A$.

Now it remains to repeat arguments from [MZ2].

Suppose that the degree of ω is $d > 0$ (the case of $d < 0$ is similar). Since $J\omega = J$, $\mathrm{Ker}\omega = (0)$, the operator ω is invertible and the degree of ω^{-1} is $-d$. Let $\mathcal{G} = J_0 + J_1 + \cdots + J_{d-1}$. We will define a new multiplication on $\mathcal{G}$.

If $a_i \in J_i$, $b_j \in J_j$, $0 \leq i, j \leq d-1$ and $i + j \leq d-1$ then $a_i * b_j = a_i b_j$, the product of the elements a_i, b_j in J. If $i + j \geq d$ then $a_i * b_j = (a_i b_j)\omega^{-1}$. It is easy to see that $(\mathcal{G}, *)$ is a $\mathbb{Z}/d\mathbb{Z}$-graded finite dimensional Jordan superalgebra and $J \simeq \mathcal{L}(\mathcal{G})$. Since the superalgebra J is graded simple it follows that the superalgebra $\mathcal{G}$ is simple. The proposition 3.1 is proved.

From now on in all cases 1,2,3 we will assume that $AD(M, M) \neq (0)$.

3.2. $A = F[t^{-n}, t^n]$

Case (a) To start with we will assume that for an arbitrary nonzero element $x \in M$ and an arbitrary $i \in \mathbb{Z}$, $xt^{ni} \neq 0$.

Let us show that this assumption implies that $R(t^{ni})$ is an invertible linear transformation on M. Indeed, let $k, l \in \mathbb{Z}$ and $k - l = 0 \mod n$. Suppose that $dim_F M_k \leq dim_F M_l$. Since $M_l(R(t^{k-l}) \subseteq M_k$ and $R(t^{k-l})$ is injective on M it follows that $dim_F M_l \leq dim_F M_k$. Hence $dim_F M_k = dim_F M_l$. This implies that $R(t^{ni}) : M \longrightarrow M$ is surjective, hence invertible.

LEMMA 3.6. *Let x, y be two homogeneous elements from M such that $AD(x, y) \neq (0)$. Then:*

(1) $t^n D(x, y) \neq 0$,

(2) $xA + yA + (xA)t^n + (yA)t^n = M$.

PROOF. Since $AD(x, y) \neq (0)$, there exists $i \in \mathbb{Z}$ such that $t^{ni}D(x, y) \neq 0$, and clearly $i \neq 0$. But $t^{ni}D(x, y) = (t^n D(x, y))it^{n(i-1)}$, which proves the first part of the lemma.

Since the linear transformation $R(t^n D(x, y)) : M \longrightarrow M$ is invertible, it follows that $M = M(t^n D(x, y)) \subseteq MD(x, y) + (MD(x, y))t^n \subseteq xA + yA + (xA)t^n + (yA)t^n$. The lemma is proved. □

The operator $U(t^n, t^{-n})$ acts on M. Let $M = \oplus M_\alpha$ be the decomposition into the sum of root spaces. This decomposition agrees with the $\mathbb{Z}$-gradation.

From $AD(M, M) \neq (0)$ it follows that there exist root vectors $x_\alpha \in M_\alpha$, $y_\beta \in M_\beta$ such that $t^n D(x_\alpha, y_\beta) \neq 0$. Then $M = M_\alpha + M_\beta$ by Lemma 3.6 (2). Hence $U(t^n, t^{-n}) : M \rightarrow M$ has no more than two eigenvalues. Lemma 3.6 (2) immediatly implies

LEMMA 3.7. *If $M = M_\alpha + M_\beta$, $\alpha \neq \beta$, and both subspaces are nonzero, then $AD(M_\alpha, M_\alpha) = AD(M_\beta, M_\beta) = (0)$.*

LEMMA 3.8. *Let $M = M_\alpha + M_\beta$, $\alpha \neq \beta$. Then $[M_\alpha, M_\beta] = (0)$.*

PROOF. Let x_α, y_β be homogeneous elements from M_α, M_β respectively and $[x_\alpha, y_\beta] = t^{ni} \neq 0$. Multiplying both sides by y_β we get $x_\alpha R(y_\beta)^2 = t^{ni} y_\beta$. The left hand side lies in M_α by Lemma 3.7 while the right hand side lies in M_β. Hence $y_\beta t^{ni} = 0$, the contradiction. The lemma is proved. □

Lemma 3.8 implies that if $M = M_\alpha + M_\beta$, $\alpha \neq \beta$,then $AD(M_\alpha, M_\beta) \subseteq [M_\alpha, M_\beta] = (0)$ and therefore $AD(M, M) = (0)$, the contradiction. Hence, there is only one root and $M = M_\alpha$.

Let $R^M < A >$ denote the subalgebra of $End_F(M)$ generated by multiplications $R(a) : M \to M$, $a \in A$. Let W be the ideal of $R^M < A >$ generated by $U(t^n, t^{-n}) - \alpha Id$. The ideal W acts nilpotently on each component M_i. Since dimensions $\dim M_i$, $i \in \mathbb{Z}$, are uniformly bounded, it follows that there exists $d \geq 1$ such that $W^d = (0)$.

LEMMA 3.9. *For arbitrary $i, j \in \mathbb{Z}$ there exists a scalar $\gamma(i,j)$ such that $R(t^{ni})R(t^{nj}) - \gamma(i,j)R(t^{n(i+j)}) \in W$.*

PROOF. Let us show that it is sufficient to consider $i, j \geq 0$. Let $j < 0$ and suppose that $\gamma(i, -j)$ exists. Then $R(t^{ni})R(t^{-nj}) - \gamma(i,-j)R((t^{n(i-j)}) \in W$. This implies

$$R(t^{ni})R(t^{-nj})U(t^{nj}) - \gamma(i,-j)R(t^{n(i-j)})U(t^{nj}) \in W.$$

Taking into account that $R(t^{nk})U(t^{nq}) = U(t^{n(k+q)}, t^{nq})$ for arbitrary $k, q \in \mathbb{Z}$, we get $R(t^{ni})R(t^{nj}) - \gamma(i,-j)U(t^{ni}, t^{nj}) \in W$.

This means that

$$R(t^{ni})R(t^{nj}) - \gamma(i,-j)(2R(t^{ni})R(t^{nj}) - R(t^{n(i+j)})) =$$

$$(1 - 2\gamma(i,-j))R(t^{ni})R(t^{nj}) + \gamma(i,-j)R(t^{n(i+j)}) \in W.$$

Remark that $\gamma(i,-j) \neq 0$ since the operator $R(t^{ni})R(t^{-nj})$ has zero kernel and thus can not lie in the nilpotent ideal W. Hence $1 - 2\gamma(i,-j) \neq 0$, for otherwise $R(t^{n(i+j)}) \in W$. Now $\gamma(i,j) = \frac{\gamma(i,-j)}{2\gamma(i,-j)-1}$.

From now on we will assume that $i \geq 0, j \geq 0$. We have $U(t^n, t^{-n}) - \alpha Id = 2R(t^n)R(t^{-n}) - (1+\alpha)Id \in W$, so $\gamma(1,-1) = \frac{1+\alpha}{2}$. Hence $\gamma(1,1) = \frac{1+\alpha}{2\alpha}$.

We will use induction on $i + j$. If $i + j \leq 2$, then the assertion is clear. Let $i + j \geq 3$, $i > 0$, $j > 0$. By the induction assuption, for any k, $0 \leq k < i + j$ there exists $\gamma(k) \in F$ such that $R(t^{nk}) = \gamma(k)R(t^n)^k \bmod W$. Hence $R(t^{ni})R(t^{nj}) = \gamma(i)\gamma(j)R(t^n)^{i+j} \bmod W$.

By the Jordan identity

$$R(t^{n(i+j)}) = 2R(t^{n(i+j-1)})R(t^n) + R(t^{n(i+j-2)})R(t^{2n}) - 2R(t^{n(i+j-2)})R(t^n)^2.$$

Hence, $R(t^{n(i+j)}) = (2\gamma(i+j-1) + \gamma(i+j-2)\gamma(2) - 2\gamma(i+j-2))R(t^n)^{i+j} \bmod W$ and, finally

$\gamma(i,j) = \frac{2\gamma(i+j-1)+\gamma(i+j-2)\gamma(2)-2\gamma(i+j-2)}{\gamma(i)\gamma(j)}$. The lemma is proved. □

REMARK 3.10. We have shown that $\gamma(k) = 2\gamma(k-1) + \gamma(k-2)\gamma(2) - 2\gamma(k-2)$. In what follows we will need the following values of γ:
$\gamma(1) = 1$, $\gamma(2) = \frac{1}{\gamma(1,1)} = \frac{2\alpha}{\alpha+1}$, $\gamma(3) = 2\gamma(2) + \gamma(1)\gamma(2) - 2\gamma(1) = \frac{4\alpha-2}{\alpha+1}$,
$\gamma(2,2) = \frac{2\gamma(3)+\gamma(2)\gamma(2)-2\gamma(2)}{\gamma(2)\gamma(2)}$.

LEMMA 3.11. *M is not a sum of two proper graded A-subbimodules.*

PROOF. Suppose that $M = M' + M''$, where M', M'' are proper graded submodules of M. From Lemma 3.7 it follows that $AD(M', M') = AD(M'', M'') = (0)$.

Choose arbitrary nonzero homogeneous $x \in M'$, $y \in M''$. We have $(yt^n)R(x)^2 = (yR(x)^2)t^n$ or equivalently, $[yt^n, x]x = (x[y,x])t^n$.

Let $\deg[y,x] = ni$. Then $(x[y,x])t^n = \gamma(i,1)x([y,x]t^n) \mod xW$. Combining this with the previous equality we get $x([yt^n, x] - \gamma(i,1)[y,x]t^n) \in xW$.

Let $[yt^n, x] - \gamma(i,1)[y,x]t^n = \xi t^{n(i+1)}$, $\xi \in F$ and suppose that $0 \neq \xi$. Then $xR(t^{n(i+1)}) \in xW$ which implies $x \in xWR(t^{n(i+1)})^{-1}$ and $x \in xW^d(R(t^{n(i+1)})^{-1})^d = (0)$, the contradiction. Hence $\xi = 0$, $[yt^n, x] = \gamma(i,1)[y,x]t^n$.

Exchanging x and y we get $[xt^n, y] = \gamma(i,1)[x,y]t^n$, hence, $[yt^n, x] = [y, xt^n]$ which means that $t^n D(x,y) = 0$. We have proved that $AD(M', M'') = (0)$ and so $AD(M, M) = (0)$, the contradiction. Lemma is proved. □

LEMMA 3.12. *Let $x, y \in M$ be elements such that $xW = (0)$, $yW \neq (0)$. Then $[x,y] = 0$.*

Remark: We do not assume that x, y are homogeneous.

PROOF. Let $[x,y] = f(t^n) \neq 0$. Then $xR(y)^2 = yf$.

Case 1. $t^n R(y)^2 = g(t^n) \neq 0$.

Let $\hat{J} = \hat{A} + \hat{M}$ be the completion of the superalgebra J. Since the operator $R(g(t^n)) : \hat{M} \to \hat{M}$ is invertible it follows that $\hat{M} = \hat{M}g = y\hat{A} + (y\hat{A})t^n$. Let $x = ya + (yb)t^n$, where $a, b \in \hat{A}$. Let $b = \sum_i b_{in}$ be the sum (may be infinite) of homogeneous components. We have

$$R(b)R(t^n) = \sum_i R(b_{in})R(t^n) = \sum_i (\gamma(i,1)R(b_{in}t^n) + \omega_{(i+1)n}),$$

where $\omega_{(i+1)n}$ is a homogeneous operator of degree $(i+1)n$ from W. Let $b' = \sum_i \gamma(i,1)b_{in}t^n$, $\omega = \sum_i \omega_{(i+1)n}$. Then $x = y(a+b') + y\omega$. This implies that $a + b' = 0$. Indeed, choose $s \geq 2$ such that $yW^s = (0)$, $yW^{s-1} \neq (0)$. Applying W^{s-1} to both sides of the expression for x, we get $(yW^{s-1})(a+b') = (0)$, which implies $a + b' = 0$. Hence, $x = y\omega$.

We have $xR(y)^2 = yf$. Hence $y = xR(y)^2R(f)^{-1}$, where $R(f)^{-1}$ lies in the completion of the algebra $R^M < A >$. Denote $D = R(y)^2$. We have $y = y\omega DR(f)^{-1} = y[\omega, D]R(f)^{-1} = y([\omega, D]R(f)^{-1})^r = y[\omega, D]^r(R(f)^{-1})^r$ for any $r \geq 1$.

Now, $[[\omega^2, D], D] = \omega^2 D^2 + D^2\omega^2 - 2D\omega^2 D$.

Since $yD = 0$ it follows that $y[[\omega^2, D], D] = y\omega^2 D^2 \in xWD^2 = (0)$. On the other hand $[[\omega^2, D], D] = 2\omega[[\omega, D], D] + 2[\omega, D]^2$.

Hence, $y[\omega, D]^2 = -y\omega[[\omega, D], D]$ and $y[\omega, D]^4 = y\omega^2[[\omega, D], D]^2 = 0$, the contradiction.

Case 2. If $t^n R(y)^2 = 0$, then $xR(y)^2 = yf$ leads to an immediate contradiction since it implies that $(yW)f = 0$. Lemma is proved. □

LEMMA 3.13. *$M(U(t^n, t^{-n}) - \alpha Id) = (0)$*

PROOF. Denote $M' = ker(U(t^n, t^{-n}) - \alpha Id)$. Our aim is to prove that $M' = M$. Suppose the contrary and let $y \in M - M'$. Choose an arbitrary element $x \in M$. There exist infinitely many scalars $\beta \in F$ such that $(x + \beta y)(U(t^n, t^{-n}) - \alpha Id) \neq 0$.

By Lemma 3.2.6 $[M', y] = [M', x + \beta y] = (0)$. Hence $[M', x] = (0)$ and therefore $[M', M] = (0)$. This implies that $M' \lhd J$, the contradiction. Lemma is proved. □

Notice that we have proved that $W = 0$

LEMMA 3.14. *Either $M = \sum_{i\in\mathbb{Z}} M_{ni}$ and $dim M_{ni} = 1$ for all $i \in \mathbb{Z}$ or n is even and $M = \sum_{i\in\mathbb{Z}} M_{n/2+ni}$, $dim M_{n/2+ni} = 1$ for all $i \in \mathbb{Z}$.*

PROOF. From Lema 3.11 it follows that there exists a homogeneous element $y \in M$ such that $M = yR^M < A >$.

Now Lemma 3.9 and $W = (0)$ imply that $M = yA$.

Let $i = deg(y)$. If $2i$ is not divisible by n then $[M, M] = (0)$. Let $2i = kn$. Taking $yt^{-[\frac{k}{2}]n}$ instead of y we can always assume that $i = 0$, which corresponds to the first case, or $i = \frac{n}{2}$, which corresponds to the second case. Lemma is proved. □

COROLLARY 3.15. *$M_j = M_i t^{n \cdot \frac{j-i}{n}}$.*

LEMMA 3.16. *(1) If $i \neq j$, $0 \neq x_i \in M_i$, $0 \neq x_j \in M_j$, then $[x_i, x_j] \neq 0$,*
(2) if $0 \neq x_i \in M_i$, then $t^n R(x_i)^2 \neq 0$.

PROOF. Suppose that $[x_i, x_j] = 0$, $j - i = kn$. Then $[x_i, x_i \cdot t^{kn}] = 0$, which implies $AR(x_i)^2 = (0)$ and so $[x_i, x_i \cdot A] = [x_i, M] = (0)$.

We have proved that $[x_i, x_j] = 0$ implies $[x_i, M] = (0)$. Hence, it implies also that $[M, M] = (0)$, the contradiction. Lemma is proved. □

LEMMA 3.17. *$\alpha = 1$*

PROOF. We have $R(t^n)R(t^n) - \gamma R(t^{2n}) = 0$, where $\gamma = \gamma(1, 1) = \frac{1+\alpha}{2\alpha}$. Suppose at first that $M = \sum_{i\in\mathbb{Z}} M_{n/2+ni}$. We can scale an element $0 \neq x_{-\frac{n}{2}} \in M_{-\frac{n}{2}}$ so that $t^n R(x_{-\frac{n}{2}})^2 = 1$. Apply the derivation $R(x_{-\frac{n}{2}})^2$ to $x_{-\frac{n}{2}}(R(t^n)R(t^n) - \gamma R(t^{2n})) = 0$. We will get $x_{-\frac{n}{2}}(2R(t^n) - 2\gamma R(t^n)) = 0$, which implies $\gamma = 1$ and therefore $\alpha = 1$.

Now let $M = \sum_{i\in\mathbb{Z}} M_{ni}$. We can scale x_{-n} so that $t^{2n} R(x_{-n})^2 = 1$. Apply $R(x_{-n})^2$ to $x_{-n}(R(t^{2n})R(t^{2n}) - \gamma(2,2)R(t^{4n})) = 0$. We will get $\gamma(2, 2) = 1$. By the remark after Lemma 3.2.4, $\gamma(2, 2) = \frac{2\gamma(3)+\gamma(2)^2-2\gamma(2)}{\gamma(2)^2} = 1$. Hence $\gamma(2) = \gamma(3)$, $\frac{2\alpha}{\alpha+1} = \frac{4\alpha-2}{\alpha+1}$, that is, $\alpha = 1$. Lemma is proved. □

COROLLARY 3.18. Using the recurrent formula for $\gamma(i,j)$ we see that $\gamma(i,j) = 1$ for all i, j.

Now scale x_0 (or $x_{-\frac{n}{2}}$) so that $[x_0t^n, x_0] = t^n$ (resp. $[x_{-\frac{n}{2}}t^n, x_{-\frac{n}{2}}] = 1$). Then it is easy to check that $[x_0t^{ni}, x_0t^{nj}] = (i-j)t^{n(i+j)}$ (resp. $[x_{-\frac{n}{2}}t^{ni}, x_{-\frac{n}{2}}t^{nj}] = (i-j)t^{n(i+j-1)}$).

Case (b) Now we will drop the assumption that for an arbitrary $i \in \mathbb{Z}$ $\mathrm{Ker}(R(t^{ni}) : M \to M) = (0)$.

Let $q = t^{ni}$, $i \neq 0$ and suppose that the kernel of $R(q) : M \to M$ is nonzero. Let $M = \oplus M_\alpha$ be the root decomposition of M with respect to $U(q^{-1}, q)$.

LEMMA 3.19. *If there is more than one root, then for an arbitrary root α we have $AD(M_\alpha, M_\alpha) = (0)$.*

PROOF. Suppose that, contrary to our assertion, there exist homogeneous elements $x, y \in M_\alpha$ such that $t^nD(x,y) = t^{nj}$. Then $t^{2n}D(x,y) = 2t^{n(j+1)}$. Choose an arbitrary element $0 \neq z \in M_\beta$, $\beta \neq \alpha$.

If $zt^{nj} = zt^{n(j+1)} = 0$, then $D(z, t^n) = D(z, t^{n(j+1)}t^{-nj}) = 0$. Hence, $zt^{2nj} = (\cdots(zt^{nj})t^n)t^n)\cdots)t^n = 0$ and therefore $zU(t^{nj}) = 0$, the contradiction.

Hence, there exists k (in fact $k = 1$ or $k = 2$) such that $t^{nk}D(x,y) = \xi t^{nl}$, $0 \neq \xi \in F$ and $zt^{nl} \neq 0$. We have $zt^{nl} \in M_\beta A \subseteq M_\beta$. On the other hand, $\xi zt^{nl} = z(t^{nk}D(x,y)) = (zr^{nk})D(x,y) - (zD(x,y))t^{nk} \in M_\alpha$, the contradiction. Lemma is proved. □

LEMMA 3.20. *If $\alpha \neq \beta$, then $AD(M_\alpha, M_\beta) = (0)$.*

PROOF. Suppose that $x \in M_\alpha$, $y \in M_\beta$ and $t^nD(x,y) = t^{nj} \neq 0$. Then an arbitrary power t^{nk} (except when $k = j-1$) lies in $AD(x,y)$. Indeed, let $l = k - j + 1 \neq 0$. Then $t^{nl}D(x,y) = lt^{n(l-1+j)} = lt^{nk}$.

Recall that $q = t^{ni}$, where $i \neq 0$. Both i and $-i$ can not be equal to $j-1$. Hence $q \in AD(x,y)$ or $q^{-1} \in AD(x,y)$.

Suppose, for example, that $q^{-1} \in AD(x,y)$. Then there exist homogeneous elements $x' \in M_\alpha$, $y' \in M_\beta$ such that $[x', y'] = q^{-1}$. Clearly, $\alpha \neq -1$ or $\beta \neq -1$.

Let $\beta \neq -1$. We claim that $y'q^{-1} \neq 0$. Indeed, if $y'q^{-1} = 0$, then $y'U(q, q^{-1}) = y'(2R(q^{-1})R(q) - 1) = -y'$, so y' belongs to the eigenvalue -1, the contradiction.

Now, $y'q^{-1} \in M_\beta$ and at the same time $y'q^{-1} = x'R(y')^2 \in M_\alpha$ by Lemma 3.19, the contradiction. Lemma is proved. □

From Lemmas 3.19 and 3.20 it follows that there is only one root. Since $\ker(R(q) : M \to M)$ lies in the eigenspace of $U(q^{-1}, q)$ corresponding to the eigenvalue -1, we conclude that the only eigenvalue of $U(q, q^{-1})$ is -1.

Hence, there exists $r \geq 1$ such that $M(U(q, q^{-1}) + 1)^r = M(R(q^{-1})R(q))^r = (0)$. We have $R(q^{-1})U(q) = R(q)$. Hence $MR(q)^{2r} = MR(q)^rR(q^{-1})^rU(q)^r = (0)$. We have proved the following assertion:

LEMMA 3.21. *If $q = t^{ni}$ and $Ker(R(q) : M \to M) \neq (0)$, then $R(q)$ acts nilpotently on M.*

Let $S = \{i \in \mathbb{Z} | R(t^{ni}) \text{ acts nilpotently on } M\}$.

From the Jordan identity it follows that for arbitrary $i, j, k \in S$ we have $\pm i \pm j \pm k \in S$. Let $\mathbb{Z}_{odd}$ denote the set of all odd integers.

LEMMA 3.22. *There exists an integer $m \in \mathbb{Z}$, $m \geq 0$, such that $S = m\mathbb{Z}_{odd}$.*

PROOF. As we have noticed above for arbitrary numbers $i, j, k \in S$ all numbers $\pm i \pm j \pm k$ also lie in S. Hence, $H = \pm S \pm S$ is a subgroup of $\mathbb{Z}$. Hence, there exists $l \geq 0$ such that $H = l\mathbb{Z}$. For an arbitrary $m \in S$ we have $S = m + H$. Since zero does not lie in S, we can assume that $1 \leq m \leq l - 1$. Then $3m = m + m + m = m + lk$. Hence, $2m$ is divisible by l, which is possible only if $2m = l$. Thus $S = m + 2m\mathbb{Z} = m\mathbb{Z}_{odd}$. Lemma is proved. □

Denote $t^{nS} = \{t^{ns} | s \in S\}$.

LEMMA 3.23. *For arbitrary homogeneous elements $x, y \in M$ we have $t^{nS}D(x, y) \subseteq Ft^{nS}$.*

PROOF. Let $a = t^{ns}$, $s \in S$, $D = D(x, y)$. We need to prove that $R(aD) = [R(a), D]$ acts nilpotently on M. But in a commutative algebra over a field of zero characteristic (the algebra $R^M < A >$ is commutative) derivations map nilpotent elements to nilpotent elements. Lemma is proved. □

LEMMA 3.24. $[M, Mt^{nS}] \subseteq Ft^{nS}$.

PROOF. Choose arbitrary homogeneous elements $x, y \in M$. Let $i \in S$. We need to prove that $[x, yt^{ni}] = 0$ or $\frac{1}{n}(deg(x) + deg(y) + ni) \in S$. By the Jordan identity the algebra $< R(t^{nS}) >$ lies in $R(t^{nm})R^M < A >$. Hence, if $MR(t^{nm})^k = (0)$ then $M(R < t^{nS} >)^k = (0)$. Thus, t^{nS} acts nilpotently on M. Hence, for a sufficiently large l we have $[M, (\cdots(M \underbrace{t^{nS})t^{nS}) \cdots)t^{nS}}_{l}] \subseteq Ft^{nS}$, because for a sufficiently large l the left hand side is equal to (0). We will show that if $l \geq 2$ then $[M, MR(t^{nS})^l] \subseteq Ft^{nS}$ implies $[M, MR(t^{nS})^{l-1}] \subseteq Ft^{nS}$.

Indeed, if $s_1, \cdots, s_{l-1} \in S$ and $[x, yR(t^{ns_1}) \cdots R(t^{ns_{l-1}})] = t^{nj}$, $j \notin S$, then $xt^{nj} = -yR(t^{ns_1}) \cdots R(t^{ns_{l-1}})R(x)^2 \subseteq MR(t^{nS})^{l-1}$ by Lemma 3.23.

Since $j \notin S$ the operator $R(t^{nj}) : M \to M$ is invertible. Hence $x \in MR(t^{nS})^{l-1}$ and $[x, MR(t^{nS})^{l-1}] \subseteq [MR(t^{nS})^{l-1}, MR(t^{nS})^{l-1}]$.

By Lemma 3.23 $[MR(t^{nS})^{l-1}, MR(t^{nS})^{l-1}] \subseteq [M, MR(t^{nS})^{2(l-1)}] + Ft^{nS} = Ft^{nS}$, because $2(l-1) \geq l$. Lemma is proved. □

LEMMA 3.25. *$m = 1$, and so, $S = \mathbb{Z}_{odd}$.*

PROOF. Let us assume that $m \geq 2$. We will prove that in this case $[M, Mt^{nS}] = (0)$.

Indeed, choose homogeneous elements $x, y \in M$ and let $s \in S$. By Lemma 3.24 if $[x, yt^{ns}] \neq 0$, then the degree of $[x, yt^{ns}]$ lies in $nm\mathbb{Z}_{odd}$. Let $[x, yt^{ns}] = t^{nmj}$, $j \in \mathbb{Z}_{odd}$.

Since $m \geq 2$ it follows that $1 \notin S$ and therefore the operator $R(t^n) : M \to M$ is invertible, so $x = x't^n$. We have

$$x'R(t^n)R(yt^{ns})R(t^n) = t^{n(mj+1)} = \frac{1}{2}x'(-R((yt^{ns})t^{2n}) + 2R(t^n)R((yt^{ns}t^n) + R(yt^{ns})R(t^{2n})).$$

The degrees of the elements $[x', (yt^{ns})t^{2n}]$, $[x't^n, (yt^{ns})t^n]$, $[x', yt^{ns}]$ are equal to $n(mj+1)$, $n(mj+1)$ and $n(mj-1)$ respectively. Hence, by Lemma 3.24 all these elements are equal to 0. This proves that $[x, yt^{ns}] = 0$ and $[M, Mt^{ns}] = (0)$.

Now Mt^{nS} is an ideal in J. Hence, $Mt^{nS} = (0)$. This implies that $t^{nm}D(M, M) = (0)$ and therefore $AD(M, M) = (0)$, the contradiction. Lemma is proved. □

LEMMA 3.26. $MR(t^n)^2 = (0)$.

PROOF. In the proof of Lemma 3.25 we have shown that $[MR(t^n)R(t^{nS}), M] = (0)$. Still let's do it again for the sake of completeness.

Choose arbitrary homogeneous elements $x, y \in M$. If $[(x.t^n)t^n, y] \neq 0$ then we can assume that $[(x.t^n)t^n, y] = t^{nj}$, $j \in Z_{odd}$. Then $(xt^n)R(t^n)R(y)R(t^n) = t^{n(j+1)} = \frac{1}{2}(xt^n)(-R(yt^{2n}) + 2R(t^n)R(yt^n) + R(y)R(t^2n)) = -\frac{1}{2}[xt^n, yt^{2n}] + [(xt^n)t^n, yt^n] + \frac{1}{2}[xt^n, y]t^{2n}$.

By Lemma 3.24, the elements $[xt^n, yt^{2n}]$, $[(xt^n)t^n, yt^n]$, $[xt^n, y]$ lie in $Ft^{n\mathbb{Z}_{odd}}$, but their degress are equal to $n(j+1)$, $n(j+1)$ and $n(j-1)$ respectively. Hence these elements are equal to 0.

We proved that $[(Mt^n)t^n, M] = (0)$. This implies that $(Mt^n)t^n$ is an ideal in J. Hence $(Mt^n)t^n = (0)$. Lemma is proved. □

LEMMA 3.27. $AD(M, Mt^n) = (0)$.

PROOF. By Lemma 3.24 $AD(M, Mt^n) \subseteq [M, Mt^n] \subseteq Ft^{n\mathbb{Z}_{odd}}$.

Let $x, y \in M$ be arbitrary homogeneous elements. If $deg(x) + deg(y) \notin n\mathbb{Z}$ then $AD(x, yt^n) = (0)$. Suppose that $deg(x) + deg(y) \in n\mathbb{Z}$. Then there exists $i \in \mathbb{Z}$ such that $ni + deg(x) + deg(y) + n \in 2n\mathbb{Z}$. Hence $t^{ni}D(x, yt^n) = 0$. Hence $AD(x, yt^n) = (0)$. Lemma is proved. □

COROLLARY 3.28. $[Mt^n, Mt^n] = (0)$.

LEMMA 3.29. *There exists a homogeneous element $x \in M$ such that $AR(x)^2 \neq (0)$.*

PROOF. Denote $A_{even} = Ft^{2n\mathbb{Z}}$, $A_{odd} = Ft^{n\mathbb{Z}_{odd}}$, $A = A_{even} + A_{odd}$.

There exist homogeneous elements $x, y \in M$ such that $AD(x, y) \neq (0)$. Hence there is $k \in \mathbb{Z}$ such that $0 \neq t^{nk}D(x, y) \in A_{even}$. Hence, $M = M(t^{nk}D(x, y)) = xA + yA + (xA)t^{nk} + (yA)t^{nk}$. By Lemma 3.26, $(MA_{odd})A_{odd} = (0)$. Hence $D(MA_{odd}, A_{even}) = D(MA_{odd}, A_{odd}A_{odd}) = (0)$. Hence, for arbitrary elements $u \in M$, $a, b \in A_{even}$ we have: $((ut^n)a)b = (ut^n)(ab)$.

This implies that $(ut^n)R(A_{even})R(A_{even}) \subseteq (ut^n)R(A_{even})$. Let us show that $uA_{odd} = (ut^n)A_{even}$. For an odd integer $k \geq 1$ we have $R(t^{nk}) = \sum \pm R(t^n)^i R(t^{2n})^j$ (see [ZSSS]), where $i \neq j$. If $i \geq 2$, then $uR(t^n)^i = 0$.

Hence $ut^{nk} = (ut^n)(\sum \pm R(t^{2n})^j) \in (ut^n)A_{even}$.

As a result we get $M = (xA_{even})A_{even} + (xt^n)A_{even} + (yA_{even})A_{even} + (yt^n)A_{even}$.

Let $i = deg(x)$, $j = deg(y)$. From $t^n D(x,y) \neq 0$ it follows that $i + j \in n\mathbb{Z}$. Moreover, by Lemma 3.23 we have $t^n D(x,y) \in Ft^{n\mathbb{Z}_{odd}}$ and therefore $i + j \in 2n\mathbb{Z}$.

The sum $J' = A_{even} + (xA_{even})A_{even} + (yA_{even})A_{even}$ is a subalgebra of J, one of those discussed in case (a).

Indeed, for an arbitrary nonzero element $0 \neq a \in A_{even}$ the operator $R(a) : M \to M$ has zero kernel and, besides, $t^{2n}D(x,y) \neq 0$. We don't know if J' is graded simple. But since A_{even} is graded simple, a proper ideal of J' can lie only in $J' \cap M$. Let $K \subseteq J' \cap M$ be the largest A_{even}-subbimodule of J' such that $[K, J' \cap M] = (0)$. The quotient algebra $\bar{J'} = J'/K$ is graded simple. Hence there exists a homogeneous element $\bar{u} \in J'_{\bar{1}}$ such that $A_{even}R(\bar{u})^2 \neq (0)$. Hence $A_{even}R(u)^2 \neq (0)$. Lemma is proved. □

Let x be a homogeneous element of M such that $t^n R(x)^2 \neq 0$. As we have seen in the proof of Lemma 3.29 (with $x = y$), $M = xA_{even} + (xt^n)A_{even}$.

The element x was constructed so that $A_{even}R(x)^2 \subseteq A_{even}$. Hence $2deg(x) \in 2n\mathbb{Z}$, $deg(x) \in n\mathbb{Z}$.

The subspace $A_{even} + xA_{even}$ is a subalgebra of J. It is simple and satisfies the assumption (a). From the classification in the case (a) it follows that for arbitrary elements $a, b \in A_{even}$ we have $(xa)b = x(ab)$.

This implies that $\Gamma = A_{even} + (xt^n)A_{even}$ is an associative commutative superalgebra. We can make the degree of x be equal to 0 or to $-n$. If $deg(x) = k \in \{0, -n\}$, then

$$\Gamma \simeq F[t^{-2n}, t^{2n}, \xi_{k+n}],$$

where ξ_{k+n} is a Grassmann variable of degree $k + n$, ${\xi_{k+n}}^2 = 0$.

For an arbitrary element $(xt^n)t^{2nq}$ we have $[(xt^n)t^{2nq}, x] \neq 0$. Otherwise $((xt^n)t^{2nq})R(x)^2 = 0$. Let $k = 0$ and let $t^n R(x)^2 = t^n$. Then $((xt^n)t^{2nq})R(x)^2 = (2q+1)(xt^n)t^{2nq} \neq 0$. If $k = -n$ and $t^n R(x)^2 = t^{-n}$ then $((xt^n)t^{2nq})R(x)^2 = (2q-1)(xt^n)t^{2n(q-1)} \neq 0$.

This implies that $J = \Gamma + \Gamma x$. The bracket on Γ is defined via $\{a, b\} = (-1)^{|b|} ax \cdot bx$. This bracket has degree $2k$.

Normalizing variables we can assume that for $k = 0$ $D = 2\frac{\partial}{\partial t^{2n}}$, $\{\xi, t^{2n}\} = \xi t^{2n}$, $\{\xi, \xi\} = -t^{2n}$;
for $k = -n$ $D = 2\frac{\partial}{\partial t^{2n}}$, $\{\xi, t^{2n}\} = -\xi$, $\{\xi, \xi\} = -t^{-2n}$.

The following proposition summarizes what we did in this section

Proposition 3.30. Let $A = F[t^{-n}, t^n]$. If $AD(M, M) = (0)$, then J is a loop superalgebra. Suppose that $AD(M, M) \neq (0)$. Then either

(1) J is isomorphic to the Kantor double $F[t^{-n}, t^n] + F[t^{-n}, t^n]x_0$ with the bracket $\{f, g\} = t^n(\frac{\partial f}{\partial t^n} g - f\frac{\partial g}{\partial t^n})$ or,

(2) n is even and $J \simeq F[t^{-n}, t^n] + F[t^{-n}, t^n]x_{-n/2}$, $\{f, g\} = \frac{\partial f}{\partial t^n} g - f\frac{\partial g}{\partial t^n}$ or

(3) $J \simeq F[t^{-2n}, t^{2n}, \xi_n] + F[t^{-2n}, t^{2n}, \xi_n]x_0$ with $\{f, g\} = 2t^{2n}(\frac{\partial f}{\partial t^{2n}} g - f\frac{\partial g}{\partial t^{2n}}) - 3t^{2n}\xi(\frac{\partial f}{\partial t^{2n}}\frac{\partial g}{\partial \xi} - \frac{\partial f}{\partial \xi}\frac{\partial g}{\partial t^{2n}}) + (-1)^{|f|}\frac{\partial f}{\partial \xi_n}\frac{\partial g}{\partial \xi_n}t^{2n}$ or

(4) $J \simeq F[t^{-2n}, t^{2n}, \xi_0] + F[t^{-2n}, t^{2n}, \xi_0]x_{-n}$, with $\{f, g\} = 2(\frac{\partial f}{\partial t^{2n}} g - f\frac{\partial g}{\partial t^{2n}}) - \xi(\frac{\partial f}{\partial t^{2n}}\frac{\partial g}{\partial \xi} - \frac{\partial f}{\partial \xi}\frac{\partial g}{\partial t^{2n}}) + (-1)^{|f|}\frac{\partial f}{\partial \xi_0}\frac{\partial g}{\partial \xi_0}t^{-2n}$.

3.3. $A = F[t_1^{-n_1}, t_1^{n_1}] \oplus F[t_2^{-n_2}, t_2^{n_2}]$

Consider $e_1 = t_1^{-n_1}t_1^{n_1}$, $e_2 = t_2^{-n_2}t_2^{n_2}$ two orthogonal idempotents, $e_1 + e_2 = 1$. Our aim in this section is to prove the following proposition. We will do that in a slightly more general form than in other sections, not assuming J to be graded simple.

PROPOSITION 3.31. If $J = A + M$ is a unital graded superalgebra (not necessarily graded simple) and $A = F[t_1^{-n_1}, t_1^{n_1}] \oplus F[t_2^{-n_2}, t_2^{n_2}]$ then $AD(M, M) = (0)$ or $J = JU(e_1) \oplus JU(e_2)$ or $n = n_1 = n_2$, $J \simeq F[t^{-n}, t^n, \xi_0] + F[t^{-n}, t^n, \xi_0]x_{-n}$ with $\{f, g\} = \frac{1}{2}(\frac{\partial f}{\partial t^n} g - f\frac{\partial g}{\partial t^n})t^{-n} + \frac{1}{4}(-1)^{|f|}\frac{\partial f}{\partial \xi_0}\frac{\partial g}{\partial \xi_0}t^{-2n}$, or $n = n_1 = n_2$ is even, $J \simeq F[t^{-n}, t^n, \xi_{n/2}] + F[t^{-n}, t^n, \xi_{n/2}]x_{-n/2}$, with $\{f, g\} = (\frac{\partial f}{\partial t^n} g - f\frac{\partial g}{\partial t^n}) - \xi(\frac{\partial f}{\partial t^{2n}}\frac{\partial g}{\partial \xi} - \frac{\partial f}{\partial \xi}\frac{\partial g}{\partial t^{2n}}) + (-1)^{|f|}\frac{\partial f}{\partial \xi_{n/2}}\frac{\partial g}{\partial \xi_{n/2}}$.

Let $M = M_1 + M_{1/2} + M_2$ be the Peirce decomposition of M with respect to e_1, e_2. If $M_{1/2} = (0)$, then $J = JU(e_1) \oplus JU(e_2)$.

Suppose that $M_{1/2} \neq (0)$. Since $M_{1/2}$ is a special bimodule over $F[t_j^{-n_j}, t_j^{n_j}]$ it follows that for arbitrary k the operator $R(t_j^{n_jk}$: $M_{1/2} \longrightarrow M_{1/2}$ is inversible with the inverse $4R(t_j^{-n_jk}$: $M_{1/2} \longrightarrow M_{1/2}$. We claim that $AD(M_i, M_i) = (0)$ for $i = 1, 2$. Indeed, if $t_j^{n_j}D(M_i, M_i) \neq (0)$, then $M_{1/2} = M_{1/2}(t_j^{n_j}D(M_i, M_i)) \subseteq ([M_{1/2}, M_i]M_i)t_j^{n_j} + ([M_{1/2}t_j^{n_j}, M_i]M_i = (0)$, because $[M_{1/2}, M_i] = (0)$. Hence, if $AD(M, M) \neq (0)$, then $AD(M_{1/2}, M_{1/2}) \neq (0)$.

Let us show that $M_i(AD(M_{1/2}, M_{1/2})) = (0)$ for $i = 1, 2$.
Indeed, $M_iA \subseteq M_i$ and therefore

$$M_i(AD(M_{1/2}, M_{1/2})) \subseteq [M_iA, M_{1/2}]M_{1/2} + ([M_i, M_{1/2}]M_{1/2})A = (0).$$

Let $A_i = F[t_i^{-n_i}, t_i^{n_i}]$. Suppose that $t_i^{n_i}D(M_{1/2}, M_{1/2}) \neq (0)$. Then the vector space $A_iD(M_{1/2}, M_{1/2})$ contains all powers $t_i^{n_ik}$, $k \in \mathbb{Z}$, except, may be, one (see the previous section). This implies that M_i annihilates $t_i^{n_i}$ and $M_iU(t_i^{n_i}) = (0)$, the contradiction. Thus, $t_i^{n_i}D(M_{1/2}, M_{1/2}) \neq (0)$ implies $M_i = 0$.

Suppose that Proposition 3.31 is valid for the subalgebra $A + M_{1/2}$ of J. Then it follows that $t_1^{n_1}D(M_{1/2}, M_{1/2}) \neq (0)$ and $t_2^{n_2}D(M_{1/2}, M_{1/2}) \neq (0)$ and therefore $M_1 = M_2 = (0)$.

We have proved that it is sufficient to prove Proposition 3.31 for the subsuperalgebra $A + M_{1/2}$. From now on, we will assume that $M = M_{1/2}$.

Let m be the smallest common multiple of n_1 and n_2.

The operator $R(t_1^m)R(t_2^m)^{-1}$ has degree 0. Hence M decomposes into a direct sum of root spaces $M = \oplus_\alpha M_\alpha$ with respect to this operator. If the decomposition contains more than one summand then $AD(M_\alpha, M_\alpha) = (0)$. Indeed, if $t_i^{n_i} D(M_\alpha, M_\alpha) \neq (0)$ then $M = M(t_i^{n_i} D(M_\alpha, M_\alpha)) \subseteq [Mt_i^{n_i}, M_\alpha]M_\alpha + ([M, M_\alpha]M_\alpha)t_i^{n_i} \subseteq M_\alpha$.

LEMMA 3.32. *If α, β are distinct roots, then $AD(M_\alpha, M_\beta) = (0)$.*

PROOF. Let $x \in M_\alpha$, $y \in M_\beta$ be two homogeneous elements. Let $[x, y] = \xi t_1^{mi} + \eta t_2^{mi} \neq 0$; $\xi, \eta \in F$, $mi = deg(x) + deg(y)$.

Then $xR(y)^2 = y(\xi t_1^{mi} + \eta t_2^{mi})$. The left hand side is killed by a power of $R(t_1^m)R(t_2^m)^{-1} - \alpha Id$ (since $AR(y)^2 = (0)$) whereas the right hand side is killed by a power of $R(t_1^m)R(t_2^m)^{-1} - \beta Id$. Hence, $y(\xi t_1^{mi} + \eta t_2^{mi}) = 0$ and, similarly $x(\xi t_1^{mi} + \eta t_2^{mi}) = 0$.

This implies $y(\xi R(t_1^m)^i + \eta R(t_2^m)^i) = x(\xi R(t_1^m)^i + \eta R(t_2^m)^i) = 0$.

On the other hand there exists $r \geq 1$ such that $y(R(t_1^m)R(t_2^m)^{-1} - \beta)^r = 0$, or equivalently $y(R(t_1^m) - \beta R(t_2^m))^r = 0$. From these two equalities it follows that $\xi\beta^i + \eta = 0$. And similarly, $\xi\alpha^i + \eta = 0$.

If $\xi = 0$, then $y(\xi t_1^{mi} + \eta t_2^{mi}) = 0$ implies that $\eta = 0$ as well contrary to our assumption.

If $\xi \neq 0$ then $\alpha^i = \beta^i$.

We have proved that $[M_\alpha, M_\beta]_{mi} \neq (0)$ implies that $\alpha^i = \beta^i$.

Let $t_j^{n_j} D((M_\alpha)_k, (M_\beta)_l) \neq (0)$, where $(M_\alpha)_k, (M_\beta)_l$ are homogeneous components of M_α, M_β respectively. Then $k+l$ is divisible by n_j. Let $k+l = n_j s$. Choose an integer $\mu \neq \frac{k+l}{m}$ and let $i = -s + \frac{m}{n_j}\mu \neq 0$. Then $t_j^{in_j} D((M_\alpha)_k, (M_\beta)_l) \neq 0$ and the degree of this subspace is $m\mu$.

Hence $\alpha^\mu = \beta^\mu$ for any $\mu \in \mathbb{Z}$ except for $\frac{k+l}{m}$. This implies $\alpha = \beta$, a contradiction. Lemma is proved. □

Hence, there is only one root, that is, there exists $\alpha \in F$, $\alpha \neq 0$, and a number $r \geq 1$ such that $M(R(t_1^m) - \alpha R(t_2^m))^r = (0)$.

Taking $\alpha^{n_2/m} t_2^{n_2}$ instead of $t_2^{n_2}$ we can assume that $\alpha = 1$, so $M(R(t_1^m) - R(t_2^m))^r = (0)$.

Let W be the ideal of $R^M < A >$ generated by $R_M(t_1^m) - R_M(t_2^m)$. Since the algebra $R^M < A >$ is commutative it follows that $W^r = (0)$. For an arbitrary polynomial f we have $R_M(f(t_1^m)) - R_M(f(t_2^m)) \in W$. Let $\Omega = \sum_{i \in \mathbb{Z}} F(t_1^{mi} - t_2^{mi})$.

As above $\hat{J} = \hat{A} + \hat{M}$ is the completion of J

LEMMA 3.33. *If an element $a \in A$ doesn't lie in Ω, then the operator $R_{\hat{M}}(a) : \hat{M} \longrightarrow \hat{M}$ is invertible.*

PROOF. Let $a = f(t_1^{n_1}) + g(t_2^{n_2})$. Let $\alpha t_1^{n_1 l_1}, \beta t_2^{n_2 l_2}$ be monomials of minimal degrees of f and g respectively. Suppose that $n_1 l_1 < n_2 l_2$. We have $R_M(\alpha t_1^{n_1 l_1})^{-1} =$

$4R_M(\alpha^{-1}t_1^{-n_1l_1})$ and $4R_M(\alpha^{-1}t_1^{-n_1l_1})R(a) = Id + \sum_{i\geq 1}\phi_i$, where ϕ_i is a linear operator of degree i. The operator $Id + \sum_{i\geq 1}\phi_i$ is invertible, and so is $R_M(a)$.

Now suppose that $n_1l_1 = n_2l_2 = ml$. The operator $R(t_1^{ml}) - R(t_2^{ml})$ is nilpotent. We have $R(a) = \alpha(R(t_1^{ml}) - R(t_2^{ml}) + (\alpha+\beta)R(t_2^{ml}) + R(f(t_1^{n_1}) - \alpha t_1^{n_1l_1} + g(t_2^{n_2}) - \beta t_2^{n_2l_2}))$. The polynomials $f' = f(t_1^{n_1}) - \alpha t_1^{n_1l_1}$, $g' = g(t_2^{n_2}) - \beta t_2^{n_2l_2}$ have less monomials than f and g respectively. If $\alpha + \beta \neq 0$ then the operator $R((\alpha+\beta)t_2^{ml} + f' + g')$ is invertible by what we proved above, and so is $R(a)$. If $\alpha + \beta = 0$ then $R(f' + g')$ is invertible by the induction assumption on the number of monomials. Lemma is proved. □

LEMMA 3.34. *Let $v \in R_M < A >$. If $v \notin W$ then $v : \hat{M} \to \hat{M}$ is invertible.*

PROOF. Let $v = \sum \alpha_{ij}R(t_1^{n_1i})R(t_2^{n_2j})$, $i, j \in \mathbb{Z}$, $\alpha_{ij} \in F$. Let $d = min\{n_1i + n_2j | \alpha_{ij} \neq 0\}$. If there is only one pair (i, j) such that $\alpha_{ij} \neq 0$ and $n_1i + n_2j = d$ then:

$$v = \alpha_{ij}R(t_1^{n_1i})R(t_2^{n_2j})(1 + \sum 16\alpha_{ij}^{-1}R(t_1^{-n_1i})R(t_2^{-n_2j})\alpha_{pq}R(t_1^{n_1p})R(t_2^{n_2q})),$$

where $pn_1 + qn_2 > in_1 + jn_2$, is invertible.

Let $n_1i + n_2j = n_1p + n_2q = d, \alpha_{ij} \neq 0, \alpha_{pq} \neq 0$. Then $n_1(i-p) = n_2(q-j) = ml$. We have $t_1^{n_1p} = t_1^{n_1i}t_1^{-ml}$, $t_2^{n_2q} = t_2^{n_2j}t_2^{ml}$. Therefore:

$$\alpha_{ij}R(t_1^{n_1i})R(t_2^{n_2j}) + \alpha_{pq}R(t_1^{n_1p})R(t_2^{n_2q}) =$$

$$\alpha_{ij}R(t_1^{n_1i})R(t_2^{n_2j}) + 4\alpha_{pq}R(t_1^{n_1i})R(t_2^{n_2j})R(t_1^{-ml})R(t_2^{ml}) =$$

$$\alpha_{ij}R(t_1^{n_1i})R(t_2^{n_2j}) + \alpha_{pq}R(t_1^{n_1i})R(t_2^{n_2j})R(t_1^{ml})^{-1}R(t_2^{ml}) =$$

$$(\alpha_{ij} + \alpha_{pq})R(t_1^{n_1i})R(t_2^{n_2j}) + \alpha_{pq}R(t_1^{n_1i})R(t_2^{n_2j})(R(t_1^{ml})^{-1}R(t_2^{ml}) - 1).$$

The second summand lies in W and therefore is nilpotent. If we replace $\alpha_{ij}R(t_1^{n_1i})R(t_2^{n_2j}) + \alpha_{pq}R(t_1^{n_1p})R(t_2^{n_2q})$ in v by $(\alpha_{ij} + \alpha_{pq})R(t_1^{n_1i})R(t_2^{n_2j})$ then the resulting operator v' is equal to v modulo W. Hence, $v' \notin W$. The operator $v' : \hat{M} \to \hat{M}$ is invertible by the induction assumption on the number of pairs (i, j) such that $\alpha_{ij} \neq 0$. Lemma is proved. □

COROLLARY 3.35. Let $\widehat{R_M(A)}$, $\hat{W}$ denote the completion of $R_M(A)$, W respectively. If $v \in \widehat{R_M(A)} - \hat{W}$, then the operator $v : \hat{M} \to \hat{M}$ is invertible.

PROOF. Let $v = \sum \alpha_{ij}R(t_1^{n_1i})R(t_2^{n_2j})$ be an infinite converging sum. Let $v_s = \sum_{n_1i+n_2j\leq s}\alpha_{ij}R(t_1^{n_1i})R(t_2^{n_2j})$. If at least for one s the operator v_s is invertible, then $v_s^{-1}v = 1 + \sum_{i\geq 1}\phi_i$ is invertible and therefore v is invertible. If for an arbitrary s, $v_s \in W$ then $v \in \hat{W}$. □

LEMMA 3.36. *Suppose that $MW^s = (0)$, $MW^{s-1} \neq (0)$, $s \geq 2$. If $x, y \in M$ are such elements that $xW^{s-1} = (0)$, $yW^{s-1} \neq 0$ then $[x, y] \in \Omega$.*

Remark. We do not assume x, y to be homogeneous.

PROOF. Suppose that $[x, y] = f \notin \Omega$. Then $R(f) : \hat{M} \to \hat{M}$ is invertible by Lemma 3.34.

Case 1 $AR(y)^2 \neq (0)$. Let $t_i^{n_i} R(y)^2 = h(t_i^{n_i}) \neq 0$. Then $\hat{M} = \hat{M}h(t_i^{n_i}) = y\hat{A} + (y\hat{A})t_i^{n_i}$. Hence $x = yv$, $v \in R^{\hat{M}} < \hat{A} >$. We claim that $v \in \hat{W}$. Indeed, applying W^{s-1} to both sides, we get $(0) = xW^{s-1} = (yW^{s-1})v$. If $v \notin \hat{W}$ then by Corollary 3.35 this implies $yW^{s-1} = (0)$, the contradiction.

Thus, $x = yv$, $v \in \hat{W}$. We have $xR(y)^2 = yf$. Since $f \notin \Omega$ we have $y = xR(y)^2 R(f)^{-1} = yvR(y^2)R(f)^{-1} = y[v, R(y)^2]R(f)^{-1} = y[v, R(y)^2]^s (R(f)^{-1})^s$ for any $s \geq 1$. As we have already mentioned above, in a commutative algebra over a field of zero characteristic a derivation maps nilpotent elements to nilpotent elements. Hence the operator $[v, R(y)^2]$ is nilpotent, the contradiction.

Case 2 $AR(y)^2 = (0)$. Then $xR(y)^2 = yf$ implies $(yW^{s-1})R(f) = (yf)W^{s-1} = xR(y)^2 W^{s-1} = xW^{s-1}R(y)^2 = (0)$. Hence, $yW^{s-1} = (0)$, the contradiction. Lemma is proved. □

COROLLARY 3.37. $[MW, M] \subseteq \Omega$

PROOF. Let s be the minimal number such that $MW^s = (0)$. If $s = 1$ then there is nothing to prove. Let $s \geq 2$. Let $y \in M$ be such an element that $yW^{s-1} \neq (0)$. For an arbitrary element $x \in M$ there exists a nonzero scalar $\beta \in F$ such that $(x + \beta y)W^{s-1} \neq (0)$. From Lemma 3.3.4 it follows that $[MW, y] \subseteq \Omega$, $[MW, x + \beta y] \subseteq \Omega$ and therefore $[MW, M] \subseteq \Omega$. □

COROLLARY 3.38. $AD(MW, M) = (0)$

PROOF. Indeed, $t_i^{n_i} D(MW, M) \subseteq F[t_i^{-n_i}, t_i^{n_i}] \cap \Omega = (0)$. □

LEMMA 3.39. $[MW, M] \neq (0)$ *(So in particular* $MW \neq (0)$*).*

PROOF. Suppose that $[MW, M] = (0)$. Then for an arbitrary element $x \in M$ we have $[t_1^m x, x] = [t_2^m x, x]$. Hence $t_1^m R(x)^2 = t_2^m R(x)^2 \in F[t_1^{-n}, t_1^n] \cap F[t_2^{-n}, t_2^n] = (0)$.

This implies that $AR(x)^2 = (0)$, contrary to our assumption. Lemma is proved. □

LEMMA 3.40. $n_1 = n_2$.

PROOF. We will show that $n_1 \neq n_2$ implies that $[MW, M] = (0)$. Choose homogeneous elements $x, y \in M$ and a homogeneous operator $\omega \in W$. Suppose that $[x\omega, y] \neq 0$. Then, by Corollary 3.37 , projections of $[x\omega, y]$ on both $F[t_1^{-n_1}, t_1^{n_1}]$, $F[t_2^{-n_2}, t_2^{n_2}]$ are not equal to 0 and $deg(x) + deg(\omega) + deg(y)$ is divisible by m. We have $y = y' t_2^{n_2}$ for some $y' \in M$, $[x\omega, y' t_2^{n_2}] = \frac{1}{2}\{x\omega, y', t_2^{n_2}\} + \frac{1}{2}\{x\omega, t_2^{n_2}, y'\}$. It is easy to see that the first summand lies in $F[t_2^{-n_2}, t_2^{n_2}]$ whereas the second summand lies in $F[t_1^{-n_1}, t_1^{n_1}]$.

Moreover, $\{x\omega, y', t_2^{n_2}\} = -t_2^{n_2}(D(x\omega, y') + R([x\omega, y'])) = -[x\omega, y']t_2^{n_2}$ by Corollary 3.38.

If $n_2 < m$ then the degree of $[x\omega, y']$ is not divisible by m. Hence $[x\omega, y'] = 0$ and $[x\omega, y]$ lies in $F[t_1^{-n_1}, t_1^{n_1}]$, the contradiction. Similarly, if $n_1 < m$ then $[MW, M] \subseteq F[t_2^{-n_2}, t_2^{n_2}]$. Lemma is proved. □

From now on we denote $n = n_1 = n_2$.

LEMMA 3.41. *M is not a sum of two proper graded A-submodules.*

PROOF. It is similar to the proof of Lemma 3.11. Let $M = M' + M''$ be a sum of proper subbimodules. Then $AD(M', M') = AD(M'', M'') = (0)$. Choose arbitrary homogeneous nonzero elements $x \in M'$, $y \in M''$. We have $[xt_1^n, y]y = ([x, y]y)t_1^n$. Let $[x, y] = \gamma t_1^{ni} + \beta t_2^{ni}$. Then $y(\gamma t_1^{ni} + \beta t_2^{ni}) = (\gamma + \beta)yt_1^{ni} + y\omega$, where $\omega = \beta R(t_2^{ni} - t_1^{ni}) \in W$. Similarly, $(y(\gamma t_1^{ni} + \beta t_2^{ni}))t_1^n = \frac{1}{2}(\gamma + \beta)yt_1^{n(i+1)} + y\tilde{\omega}$, $\tilde{\omega} \in W$.

Hence, $y([xt_1^n, y] - \frac{1}{2}(\gamma + \beta)t_1^{n(i+1)}) = y\tilde{\omega}$.

By Lemma 3.33 $[xt_1^n, y] - \frac{1}{2}(\gamma + \beta)t_1^{n(i+1)} \in \Omega$.

Exchanging x and y we get $[y, x] = -\gamma t_1^{ni} - \beta t_2^{ni}$ and therefore $[yt_1^n, x] + \frac{1}{2}(\gamma + \beta)t_1^{n(i+1)} \in \Omega$.

Hence, $t_1^n D(x, y) \in F[t_1^{-n}, t_1^n] \cap \Omega = (0)$. We have proved that $AD(M', M'') = (0)$ which implies $AD(M, M) = (0)$, the contradiction. Lemma is proved. □

LEMMA 3.42. *(1) If $t_1^n R(y)^2 \neq 0$ then $t_2^n R(y)^2 \neq 0$.*
(2) If $t_1^{ni} R(y)^2 = e_1$ then $t_2^{ni} R(y)^2 = e_2$.

PROOF. We have already mentioned that by Lemma 3.33 $\Omega R(y)^2 \subseteq \Omega$ for an arbitrary $y \in M$. This implies both assertions. Lemma is proved. □

LEMMA 3.43. *The superalgebra J is graded simple.*

(Recall that we assume $M = M_{1/2}$ and $AD(M, M) \neq (0)$.)

PROOF. Let $(0) \neq I \lhd J$. If $I \cap A \neq (0)$, then $e_1 \in I$ or $e_2 \in I$. In both cases $M \subset I$. From Lemma 3.42 (1) it follows that $[M, M]$ is not contained in $F[t_1^{-n}, t_1^n]$ or in $F[t_2^{-n}, t_2^n]$. Hence $I = J$.

Now let $I \subset M$. Then $[I, M] = (0)$. Choose arbitrary (not necessarily homogeneous) elements $y, z \in M$, $a \in A$. By the Jordan identity:

$$I(R([y, z])R(a) + R(ya)R(z) - R(za)R(y) - R(y)R(z)R(a) +$$
$$R(a)R(z)R(y) - R([ya, z])) = (0).$$

This implies $I(R([y, z])R(a) - R([ya, z])) = (0)$.

Now let $x \in M$ be such element that $AR(x)^2 \neq (0)$. Then by Lemma 3.42 (1) $t_1^n R(x)^2 \neq 0$.

Let $y = x$, $z = xt_1^n$, $a = t_1^n$. Then $[ya, z] = 0$ and therefore $IR([x, xt_1^n])R(t_1^n) = (0)$, the contradiction. Lemma is proved. □

It follows from Lemma 3.41 that there exists a homogeneous element $x \in M$ such that $M = xR < A >$.

Let as denote $A_1 = F[t_1^{-n}, t_1^n]$.

Let us show that $AR(x)^2 \neq (0)$. Indeed, $M = xA_1 + MW$. Suppose that $t_1^n R(x)^2 = 0$.

Then $A_1 D(M,M) \subseteq A_1 D(xA_1+MW, xA_1+MW) = A_1 D(xA_1, xA_1) \subseteq [xA_1, xA_1]$.

We claim that $[xt_1^{ni}, xt_1^{nj}] = 0$ for any $i, j \in \mathbb{Z}$. If $i = 0$ or $j = 0$ or $i = j = 1$ then the assertion is obvious. We have

$[xt_1^{ni}, xt_1^{nj}] = [x, (xt_1^{nj})t_1^{ni}] + t_1^{ni} D(x, xt_1^{nj})$,

$[x, (xt_1^{nj})t_1^{ni}] = \frac{1}{2}[x, xt_1^{n(i+j)}] = 0$,

$t_1^{ni} D(x, xt_1^{nj}) = it_1^{n(i-1)}(t_1^n D(x, xt_1^{nj}))$,

$t_1^n D(x, xt_1^{nj}) = [xt_1^n, xt_1^{nj}]$.

Hence it is sufficient to assume $i = 1$. Repeating the argument above we can assume $j = 1$. This proves the claim.

Since $t_1^n R(x)^2 \neq 0$, it follows that $x \in M_k$, where either $k \in n\mathbb{Z}$ or n is even and $k \in \frac{-n}{2} + n\mathbb{Z}$. Without loss of generality we will assume that $k = -n$ or $k = \frac{-n}{2}$. Moreover, we will scale x so that $t_1^n R(x)^2 = e_1$ if $x \in M_{-n/2}$ or $t_1^{2n} R(x)^2 = e_1$ if $x \in M_{-n}$.

Denote $a = t_1^n - t_2^n$.

LEMMA 3.44. $MR(a)^2 = 0$

PROOF. Since $M = xR < A >$ it is sufficient to prove that $(xa)a = 0$. Let s be the maximal number such that $xR(a)^s \neq 0$. Suppose that $s \geq 2$.

<u>Case 1</u> $k = \frac{-n}{2}$. Then $t_1^n R(x)^2 = e_1$. We will prove that $[MR(a)^s, M] = (0)$. This will imply that $MR(a)^s$ is an ideal in J, contradicting Lemma 3.43.

Since $M = xA_1 + MW$ and $MWR(a)^s = (0)$ (because $W = R(a)R^M < A >$)) it is sufficient to prove that $[xR(a)^s R(t_1^{ni}), M] = (0)$.

Furthermore, by Corollary 3.38 $[xR(a)^s R(t_1^{ni}), MW] = [xR(a)^s W R(t_1^{ni}), M] = (0)$. Now we have only to prove that $[xR(a)^s, xt_1^{ni}] = 0$ for any $i \in \mathbb{Z}$. Suppose that this is not the case. Then $[xR(a)^s, xt_1^{ni}] = \gamma(t_1^{n(s+i-1)} - t_2^{n(s+i-1)})$, $0 \neq \gamma \in F$.

Suppose first that $i \geq 0$. Apply $R(x)^2$ to both sides of the last equality. Since $aR(x)^2 = e_1 - e_2$ and $M(e_1 - e_2) = (0)$, we get $i[xR(a)^s, xt_1^{n(i-1)}] = \gamma(s + i - 1)(t_1^{n(s+i-2)} - t_2^{n(s+i-2)})$.

Proceeding in this way we can reduce i to one:

$$[xR(a)^s, x] = \xi(t_1^{n(s-1)} - t_2^{n(s-1)}),\ \xi \neq 0$$

Apply again $R(x)^2$. The left hand side becomes 0 while the right hand side remains non-zero, the contradiction.

Now let $i < 0$. We have $t_1^{-n} R(xt_1^n)^2 = [(xt_1^n)t_1^{-n}, xt_1^n] = \frac{1}{2}[xe_1, xt_1^n] = \frac{-1}{4}$.

Now, $[xR(a)^s, xt_1^{ni}] = 4[xR(a)^s, (xt_1^{n(i-2)})R(t_1^n)R(t_1^n)] = 4[(xt_1^n)R(a)^s, (xt_1^n)t_1^{n(i-2)}]$.

If this element is not equal to zero, then applying $R(xt_1^n)^2$ repeatedly (as in the case $i \geq 0$) we will get a contradiction.

Case 2 $k = -n$. Then $t_1^{2n}R(x)^2 = e_1$.
Arguing as above we see that it is sufficient to prove that $[xR(a)^s, xt_1^{ni}] = 0$.
Denote $t^n = t_1^n + t_2^n$ (remark that t_1, t_2, t do not have any meaning) and let $b = t_1^{2n} - t_2^{2n}$. We have $bR(x)^2 = e_1 - e_2$, $R(b) = 2R(t_1^n)^2 - 2R(t_2^n)^2 = 2R(a)R(t^n)$, $R(t^n) = 2R(t_1^n) - R(a)$.

Now, $[xR(a)^s, xt_1^{ni}] = 2^s[xR(a)^s, (xt_1^{n(i-s)})R(t_1^n)^s] =$
$2^s[xR(a)^s, (xt_1^{n(i-s)})(\frac{1}{2}R(t^n) + \frac{1}{2}R(a))^s] = [xR(a)^s, (xt_1^{n(i-s)})R(t^n)^s] =$
$[x(R(a)R(t^n))^s, xt_1^{n(i-s)}] = \frac{1}{2^s}[xR(b)^s, xt_1^{n(i-s)}]$.
Now we have to prove that $[xR(b)^s, xt_1^{nj}] = 0$ for any $j \in \mathbb{Z}$.
Let j be even and nonnegative. Then $[xR(b)^s, x(t_1^{2n})^{j/2}] = \gamma(t_1^{2n(s+j/2-1)} - t_2^{2n(s+j/2-1)})$ and we can apply $R(x)^2$ as above.

If j is even and negative we replace x with xt_1^{2n} and scale x so that $t_1^{-2n}R(xt_1^{2n})^2 = 1$. Then, as in the case 1, we reduce our expression to
$[(xt_1^{2n})R(b)^s, (xt_1^{2n})t_1^{n(j-4)}]$ and repeatedly apply $R(xt_1^{2n})^2$ as before.
Now let j be odd. We have $[xR(b)^s, (xt_1^{n(j-1)})t_1^n]t_1^n =$
$\frac{1}{2}xR(b)^s(-R(t_1^{2n})R(xt_1^{n(j-1)}) + R(xt_1^{n(j-1)})R(t_1^n)^2 + R(t_1^n)^2R(xt_1^{n(j-1)})+$
$R(((xt_1^{n(j-1)})t_1^n)t_1^n))$.
Each summand on the right hand side is equal to zero because of what we proved for even j. Lemma is proved. □

COROLLARY 3.45. $x(R(t^n)^2 - R(t_1^{2n} + t_2^{2n})) = 0$.

PROOF. $(R(t_1^n + t_2^n))^2 - R(t_1^{2n} + t_2^{2n}) = (R(t_1^n) + R(t_2^n))^2 - 2R(t_1^n)^2 - 2R(t_2^n)^2 = -(R(t_1^n) - R(t_2^n))^2 = -R(a)^2$, which implies the result. □

COROLLARY 3.46. $M = \sum_{i\in\mathbb{Z}} Fx.t^{ni} + \sum_{i\in\mathbb{Z}} F(xa)t^{ni}$.

PROOF. Denote $v = e_1 - e_2$. Then $A = F[t^{-n}, t^n] + vF[t^{-n}, t^n]$. The algebra $R^M < A >$ is generated by $R_M(t^{ni})$, $i \in \mathbb{Z}$, by $R_M(v) = 0$ and by $R(vt^n) = R(a)$. A nonzero product of these generators can contain no more than one $R(a)$. Moreover, $R_M(t^{ni})R_M(t^{nj}) = R_M(t^{n(i+j)})$ by Corollary 3.45. This proves the claim.
Now the subsuperalgebra $\Gamma = \sum_{i\in\mathbb{Z}} Ft^{ni} + \sum_{i\in\mathbb{Z}} F(xa)t^{ni}$ is isomorphic to $F[t^{-n}, t^n, \xi_k]$, where $k = 0$ if $deg(x) = -n$ and $k = \frac{n}{2}$ if $deg(x) = \frac{-n}{2}$ (n is even). In both cases $J = \Gamma + \Gamma x$. Proposition 3.31 is proved. □

Remark We have completely analized the case when $A = \mathcal{L}(G)$, $\mathcal{G} = F1 + V$, $dim_F V = 1$. Indeed, if V has degree 0 then $\mathcal{L}(G) \simeq F[t_1^{-n}, t_1^n] \oplus F[t_2^{-n}, t_2^n]$.
If n is even, V has degree $\frac{n}{2}$, $v \in V$ and $v^2 = 1$, then $\mathcal{L}(G) \simeq F[q^{-1}, q]$, where $q = v \otimes t^{n+2}$.

3.4. $A = \mathcal{L}(G') \oplus \mathcal{L}(G)''$

Recall that $\mathcal{G}'$, $\mathcal{G}''$ are simple finite dimensional Jordan algebras graded by finite cyclic groups. Our aim is to prove that either $Z(A)D(M,M) = (0)$ and so J is a loop superalgebra by Proposition 3.1 or $Z(A)D(M,M) \neq (0)$ but both $\mathcal{L}(G')$ and $\mathcal{L}(G'')$ are isomorphic to algebras of Laurent polynomials (see section 3).

Suppose that $A = \mathcal{L}(G') + \mathcal{L}(G'')$ and $\mathcal{L}(G')$ is not an algebra of Laurent polynomials.

LEMMA 3.47. *Let $\mathcal{G}$ be a simple Jordan F-algebra of finite dimension > 1. Let $\mathcal{G} = \mathcal{G}_0 + \mathcal{G}_1 + \cdots + \mathcal{G}_{n-1}$ be a $\mathbb{Z}/n\mathbb{Z}$-gradation on $\mathcal{G}$. Then there exist nonzero homogeneous elements $a, b \in \mathcal{G}$ such that $aU(b) = 0$.*

PROOF. If we assume the contrary then an arbitrary nonzero homogeneous element of $\mathcal{G}$ is invertible. Let $N : \mathcal{G} \to F$ be the norm of $\mathcal{G}$. An element $a \in \mathcal{G}$ is invertible if and only if $N(a) \neq 0$ (see [J3], Cor 2, p. 227). We have $N(a + xb) = N(a) + \cdots + x^d N(b)$, where d is the degree of $\mathcal{G}$. Hence, if $N(b) \neq 0$ then for an arbitrary element $a \in \mathcal{G}$ there exists a scalar $\alpha \in F$ such that $N(a + \alpha b) = 0$. (Here we use the assumption that F is algebraically closed). This implies that homogeneous spaces $\mathcal{G}_i$ have dimensions ≤ 1.

Let $1 \leq i \leq n-1$ be the minimal number such that $\mathcal{G}_i \neq (0)$ and let $0 \neq a_i \in \mathcal{G}_i$. We claim that $\mathcal{G} =< a_i >$. Indeed, if $1 \leq j \leq n-1$ and $j = ik$ then $\mathcal{G}_j = Fa_i^k$ because $0 \neq a_i^k \in \mathcal{G}_j$. Suppose that j is not a multiple of i, $0 \neq a_j \in \mathcal{G}_j$. If $1 < j \leq 2i$ then $0 \neq a_j^{-1}U(a_i) \in \mathcal{G}_{2i-j}$, $1 \leq 2i - j < i$, the contradiction.

Let $2i < j$. Then $0 \neq a_j U(a_i^{-1}) \in \mathcal{G}_{j-2i}$. By the induction assumption (about j) we have $a_j U(a_i^{-1}) \in < a_i >$, which implies $a_j \in < a_i >$.

Hence, $\mathcal{G} =< a_i >$ is associative and thus $\mathcal{G} = F1$, which contradicts our assumption. Lemma is proved. □

The completion $\widehat{\mathcal{L}(G')}$ is a simple Jordan algebra which is finite dimensional over its center. By Lemma 3.4.1 $\widehat{\mathcal{L}(G')}$ is not a division algebra. Hence, $\widehat{\mathcal{L}(G')}$ contains two orthogonal idempotents e_1', e_2' such that $e_1' + e_2' = e'$, the identity of $\mathcal{G}'$. Let e'' denote the identity of the algebra $\mathcal{G}''$. By [R-Z] we have $M = \{e', M, e''\} = \{e_1', M, e''\} + \{e_2', M, e''\}$.

Now $Z(A)D(\{e_1', M, e''\}, \{e_2', M, e''\}) \subseteq Z(A) \cap [\{e_1', M, e''\}, \{e_2', M, e''\}] = (0)$, because $[\{e_1', M, e''\}, \{e_2', M, e''\}] \subseteq \{e_1', \mathcal{L}(G'), e_2'\}$

Furthermore $Z(A)D(\{e_i', M, e''\}, \{e_i', M, e''\}) \subseteq$
$Z(A) \cap [\{e_i', M, e''\}, \{e_i', M, e''\}] \subseteq Z(A) \cap (\{e_i', \mathcal{L}(G'), e_i'\} + \mathcal{L}(G''))$, $i = 1, 2$. This implies that $Z(\mathcal{L}(G'))D(M,M) = (0)$.

We say that an action of a Lie algebra L on a module V is SPI (see [B]) if the associative algebra $< L|_V > \leq End_F(V)$ is PI.

We claim that the action of the Lie algebra $\mathcal{D} = D(M,M)$ on M is SPI.

Indeed, choose a nonzero homogeneous central element $0 \neq c_n \in Z(\mathcal{L}(G'))$, $c_n\mathcal{D} = (0)$. The operator $R_M(c_n)$ is invertible because $R_M(c_n)R_M(c_n^{-1}) = \frac{1}{4}Id_M$. Hence, $M = \sum_{0 \leq j \leq n-1, i \in \mathbb{Z}} c_n^i M_j$ and the algebra $< \mathcal{D} > |_M$ is embeddable in the algebra of $s \times s$ matrices over $< R(c_n), R(c_n^{-1}) >$, $s = dim_F(\sum_{j=0}^{n-1} M_j)$.

A. Regev [Re] proved that a tensor product of two PI-algebras is PI. This implies that if actions of a Lie algebra $\mathcal{D}$ on modules V and W are both SPI then the action of $\mathcal{D}$ on $V \otimes_F W$ is also SPI. Hence the action of $\mathcal{D}$ on $[M, M]$ is SPI.

Remark that the subspace $[M, M]$ is not contained in $\mathcal{L}(G')$. Otherwise $\mathcal{L}(G') + M$ would be an ideal in J.

Without loss of generality we can assume that $\mathcal{L}(G'')$ is the algebra of Laurent polynomials. Otherwise $Z(\mathcal{L}(G''))\mathcal{D} = (0)$ and so $Z(A)\mathcal{D} = (0)$.

Let $\mathcal{L}(G'') \simeq F[t^{-r}, t^r]$.

LEMMA 3.48. *The action of $\mathcal{D}$ on $\mathcal{L}(G'')$ is SPI.*

PROOF. Let $B = \mathcal{L}(G'')\mathcal{D}$. Since $B \neq (0)$ it follows that B contains all powers t^{rs}, $s \in \mathbb{Z}$, except, may be, one. This implies that $\mathcal{L}(G'') = B + BB$. From the inclusion $B \subseteq [M, M]$ it follows tht the action of $\mathcal{D}$ on B is SPI. Now Regev's theorem implies that the action of $\mathcal{D}$ on $\mathcal{L}(G'') = B + BB$ is SPI as well. Lemma is proved. □

LEMMA 3.49. *Consider the algebra of Laurent polynomials $F[x^{-1}, x]$ and its derivations $d_i = \frac{d}{dx}x^{i+1}$. If $0 < i < j$ then d_i and d_j generate the associative subalgebra of $End_F F[x^{-1}, x]$ which is not PI.*

PROOF. Let L be the Lie algebra generated by d_i and d_j. It is well known that L does not have finite dimensional irreducible modules of dimension > 1. Suppose that the algebra $< d_i, d_j >$ is PI. Without loss of generality we can assume that the ground field F is not countable. Let P be a primitive homomorphic image of the algebra $< d_i, d_j >$. By theorem of I. Kaplansky (see [J1],[R2]) P is isomorphic to a matrix algebra over a division algebra D. We have $dim_F D \leq dim_F < d_i, d_j >< card(F)$. By theorem of S. Amitsur ([J1],[R3]), $D = F$, so the algebra P is finite dimensional. In view of what we said about finite-dimensional irreducible L-modules we have $P = F$. Let Jac($< d_i, d_j >$) be the Jacobson radical of $< d_i, d_j >$. Since $< d_i, d_j >$/Jac($< d_i, d_j >$) is a subdirect product of primitive homomorphic images of $< d_i, d_j >$ (see [J1]) the algebra $< d_i, d_j >$/Jac($< d_i, d_j >$) is commutative. Hence $[d_i, d_j] = (i - j)d_{i+j} \in Jac(< d_i, d_j >)$. Again by the theorem of S. Amitsur ([J1], [R3]) the ideal Jac($< d_i, d_j >$) is nil. Hence, the derivation d_{i+j} is nilpotent, which is not the case. Lemma is proved. □

Remark. If $i < 0$, $j < 0$, $i \neq j$, then the algebra $< d_i, d_j >$ is also not PI.

LEMMA 3.50. *Let L be a graded subalgebra of the Lie algebra $Der(F[x^{-1}, x])$ and supose that the action of L on $F[x^{-1}, x]$ is SPI. Then L is one of the following Lie algebras: Fd_i, $Fd_0 + Fd_i$, $Fd_{-i} + Fd_0 + Fd_i$.*

PROOF. By Lemma 3.49 the algebra L does not contain derivations d_i, d_j such that $i \neq j$ and $ij > 0$.

If the only d_i that lies in L is d_0 then $L = Fd_0$.

Suppose that $d_i \in L$, $i > 0$. If $L \neq Fd_i$ then L contains an element d_j, $j \neq i$. If the only such j is 0 then $L = Fd_0 + Fd_i$. Let $j \neq 0$. By Lemma 3.49, $j < 0$. If the only such j is $-i$ then $L = Fd_{-i} + Fd_0 + Fd_i$.

Suppose that $j \neq -i$. Then $[d_i, d_j] = (i-j)d_{i+j} \in L$. If $i+j > 0$ then $< d_i, d_{i+j} >$ is not PI. If $i+j < 0$ then $< d_i, d_{i+j} >$ is not PI. In both cases we get a contradiction. Lemma is proved. □

Now we are ready to produce a contradiction (that stemms from the assumptions that J is graded simple, $AD(M, M) \neq (0)$ and $\mathcal{L}(G')$ is not isomorphic to an algebra of Laurent polynomials).

Let L denote the restriction of the action of $\mathcal{D}$ on $\mathcal{L}(G') = F[t^{-r}, t^r]$. From Lemmas 3.48 and 3.50 it follows that $dim_F L \leq 3$ and there exists $k \geq 1$ such that $F[t^{-r}, t^r]D(M_i, M_j) = (0)$ as soon as $i + j \geq k$. Let $c = t^r$ and choose an element $x \in M$ such that $cR(x)^2 \neq 0$. We do not assume x to be homogeneous, but let $x \in \sum_{j \geq s} M_j$. Choose $i \geq 1$ such that $2s + i \geq k$. Then $[xc^{2i}, x] = 2[(xc^i)c^i, x]$, since $xU(c^i) = 0$. Now $[(xc^i)c^i, x] = c^i D(xc^i, x) = 0$, the promised contradiction.

PROPOSITION 3.51. Let A be a sum of two loop algebras $A = \mathcal{L}(G') \oplus \mathcal{L}(G'')$, where $\mathcal{G}', \mathcal{G}''$ are finite dimensional simple Jordan algebras graded by finite cyclic groups. Then $AD(M, M) = (0)$, in which case J is a loop superalgebra or both $\mathcal{L}(G')$, $\mathcal{L}(G'')$ are algebras of Laurent polynomials. In the latter case the structure of J is described by Proposition 3.31.

CHAPTER 4

A is a Loop Algebra

4.1. General Results

In this chapter we will address the case when $A = \mathcal{L}(G)$ is a loop algebra corresponding to a $\mathbb{Z}/n\mathbb{Z}$-graded simple finite dimensional Jordan algebra $\mathcal{G}$.

LEMMA 4.1. *Let $\mathcal{G}$ be a simple Jordan algebra which is finite dimensional over its center Z. Let $\tilde{Z}$ be the algebraic closure of Z and suppose that $\mathcal{G} \otimes_Z \tilde{Z}$ has capacity ≥ 3. Let e, f be two orthogonal idempotents of $\mathcal{G}$ such that $e + f = 1$. If V is a $\mathcal{G}$-bimodule (over F, not over Z) such that $V = \{e, V, f\}$, then $V = (0)$.*

PROOF. Let $\mathcal{G} = \mathcal{G}_e + \mathcal{G}_{e,f} + \mathcal{G}_f$ be the decomposition of $\mathcal{G}$ with respect to e, f. Then $V \cdot \mathcal{G}_{e,f} \subseteq \{e, V, e\} + \{f, V, f\} = (0)$.

<u>Claim 1</u> $VD(Z, \mathcal{G}) = (0)$.

Let's check that $\mathcal{G} = \mathcal{G}_{e,f}^2 e + \mathcal{G}_{e,f}^2 + \mathcal{G}_{e,f}$. Since $\mathcal{G}$ is simple it is sufficient to verify that the right hand side is an ideal. It is clearly invariant with respect to $R(\mathcal{G}_{e,f})$. Choose $a_e \in \mathcal{G}_e$ and let us check that the right hand side is invariant with respect to $R(a_e)$. Choose $b_{e,f} \in \mathcal{G}_{e,f}$. By the Jordan identity

$$b_{e,f}^2 R(a_e) = b_{e,f} R(b_{e,f}) R(a_e) R(e) = b_{e,f}(-R(e)R(a_e)R(b_{e,f}) -$$
$$R((b_{e,f}e)a_e) + R(b_{e,f}a_e)R(e) + R(b_{e,f}e)R(a_e) + R(a_e e)R(b_{e,f})).$$

The first, second and fifth summands lie in $\mathcal{G}_{e,f}^2$. The third summand lies in $\mathcal{G}_{e,f}^2 e$ and the fourth summand is equal to $\frac{1}{2} b_{e,f}^2 R(a_e)$.

This implies that $\mathcal{G}_{e,f}^2 R(a_e) \subseteq \mathcal{G}_{e,f}^2 e + \mathcal{G}_{e,f}^2$.
And $\mathcal{G}_{e,f}^2 R(e) R(a_e) = \mathcal{G}_{e,f}^2 R(a_e) R(e) \subseteq \mathcal{G}_{e,f}^2 e$.
Since $\mathcal{G}_{e,f}^2 e + \mathcal{G}_{e,f}^2 + \mathcal{G}_{e,f} = \mathcal{G}_{e,f}^2 f + \mathcal{G}_{e,f}^2 + \mathcal{G}_{e,f}$ it follows that $\mathcal{G}_{e,f}^2 e + \mathcal{G}_{e,f}^2 + \mathcal{G}_{e,f}$ is invariant with respect to $R(\mathcal{G}_f)$ as well.

Now, $VD(Z, \mathcal{G}_{e,f}) \subseteq V\mathcal{G}_{e,f} + (V\mathcal{G}_{e,f})Z = (0)$ and $VD(Z, \mathcal{G}_{e,f}^2) \subseteq VD(\mathcal{G}_{e,f}, \mathcal{G}_{e,f}) = (0)$. Furthermore, $Z \subseteq Z_e + Z_f$ where $Z_e = ZU(e)$, $Z_f = ZU(f)$.
Now, $D(Z, \mathcal{G}_{e,f}^2 e) = D(Z_e, \mathcal{G}_{e,f}^2 e) \subseteq D(Z_e, \mathcal{G}_{e,f}^2) + D(Z_e \mathcal{G}_{e,f}^2, e) \subseteq D(\mathcal{G}_{e,f}, \mathcal{G}_{e,f}) + D(\mathcal{G}_e, e) = (0)$. This proves the claim.

Consider the simple $\tilde{Z}$-algebra $\tilde{\mathcal{G}} = \mathcal{G} \otimes_Z \tilde{Z}$, of capacity ≥ 3.
Let $e = \sum_{i=1}^r e_i$, $f = \sum_{j=r+1}^s e_j$, where $e_1, \ldots, e_s$ are pairwise orthogonal primitive idempotents of $\tilde{\mathcal{G}}$. From $s \geq 3$ it follows that $r \geq 2$ or $s - r \geq 2$. Let $r \geq 2$.

The idempotents e_i are pairwise strongly connected (see [J3]). For any $1 \leq i \leq r$, $r+1 \leq j \leq s$ choose an element $a_{ij} \in \{e_i, \tilde{\mathcal{G}}, e_j\}$ such that $a_{ij}^2 = e_i + e_j$. If $1 \leq k \neq i \leq r$ then $e_j = a_{ij}^2 a_{kj}^2$. Hence all idempotents $e_1, \ldots, e_s$ lie in $\tilde{\mathcal{G}}_{e,f}^2 + \tilde{\mathcal{G}}_{e,f}^2 \tilde{\mathcal{G}}_{e,f}^2$. Hence, $1 \in \tilde{\mathcal{G}}_{e,f}^2 + \tilde{\mathcal{G}}_{e,f}^2 \tilde{\mathcal{G}}_{e,f}^2$. This implies that $1 \in \tilde{\mathcal{G}}_{e,f}^2 + \tilde{(\mathcal{G}_{e,f}^2)}^2$ and $Z \subseteq \tilde{\mathcal{G}}_{e,f}^2 + (\tilde{\mathcal{G}}_{e,f}^2)^2$. Hence $Z \subseteq \mathcal{G}_{e,f}^2 + (\mathcal{G}_{e,f}^2)^2$

Claim 2. $\mathcal{G}D(V, Z) = (0)$.

Indeed, $\mathcal{G}D(V, \mathcal{G}_{e,f}^2 + (\mathcal{G}_{e,f}^2)^2) \subseteq \mathcal{G}D(V\mathcal{G}_{e,f}, \mathcal{G}_{e,f}) = (0)$.

We have proved that V is a vector space over the field Z and a $\mathcal{G}$-bimodule over Z. Hence $\tilde{V} = V \otimes_Z \tilde{Z}$ is a bimodule over $\tilde{\mathcal{G}}$.

The set H of all elements $h \in \tilde{\mathcal{G}}$ such that $(\tilde{\mathcal{G}}_e + \tilde{\mathcal{G}}_f)D(h, \tilde{V}) = (0)$ is a subalgebra of $\tilde{\mathcal{G}}$ and $\tilde{\mathcal{G}}_{e,f} \subseteq H$.

Recall that $1 \leq k \neq i \leq r$, $r+1 \leq j \leq s$. We have $a_{ij} \in H$, $a_{kj} \in H$. Hence $a_{ij}^2 = e_i + e_j$ and $a_{kj}^2 = e_k + e_j$ lie in H.

This implies that $e_j = (e_i + e_j)(e_k + e_j) \in H$ and therefore $f \in H$.

Now, for an arbitrary $v \in V$ we have $fD(v, f) = (vf)f - vf = 0$. But $v \in \{e, V, f\}$ and therefore $vf = \frac{1}{2}v$. Hence $(\frac{1}{4} - \frac{1}{2})v = 0$, that is, $v = 0$. Lemma is proved. □

LEMMA 4.2. *Let $A = \mathcal{L}(G)$ be a loop algebra corresponding to a simple finite dimensional $\mathbb{Z}/n\mathbb{Z}$-graded F-algebra $\mathcal{G}$ of capacity ≥ 3. Then $Z(A)D(M, M) = (0)$.*

PROOF. Consider the completion $\widehat{J} = \widehat{A} + \widehat{M}$ of the superalgebra J. Then $\widehat{A}$ is finite dimensional over its center $\widehat{Z} = F[[t^{-n}, t^n]]$,the field of Laurent series over F. Since A is graded simple, it follows that $\hat{A}$ is topologically simple. Hence $\hat{A}$ is prime, hence simple.

By Lemma 3.47 $\widehat{A}$ is not a division algebra. Hence $\widehat{A}$ contains a pair of orthogonal idempotents e, f such that $e + f = 1$. Let $\widehat{M} = \widehat{M}_e + \widehat{M}_{ef} + \widehat{M}_f$ be the Peirce decomposition of $\widehat{M}$.

Let $\tilde{Z}$ be the algebraic closure of $\widehat{Z}$. The algebra $\widehat{A} \otimes_{\widehat{Z}} \tilde{Z}$ satisfies the same multilinear identities with coefficients from F as the algebra $\mathcal{G}$.

In particular $\widehat{A} \otimes_{\widehat{Z}} \tilde{Z}$ does not satisfy multilinear "tetrade-eating" identities (see [Z3] or chapter 1). Hence $\widehat{A} \otimes_{\widehat{Z}} \tilde{Z}$ has capacity ≥ 3.

Let $(\widehat{M}_e + \widehat{M}_f)R < \widehat{A} >$ be the subbimodule of $\widehat{M}$ generated by $\widehat{M}_e + \widehat{M}_f$, and let $V = \widehat{M}/(\widehat{M}_e + \widehat{M}_f)R < \widehat{A} >$. By Lemma 4.1 $V = (0)$.

Hence $\widehat{M}$ is generated as an $\hat{A}$-bimodule by $\widehat{M}_e$ and $\widehat{M}_f$.

Let $\widehat{M}' = \{x \in \widehat{M} | \widehat{Z}D(x, \widehat{M}) = (0)\}$.

For arbitrary elements $x \in \widehat{M}'$, $y \in \widehat{M}$, $a \in \widehat{A}$ we have $D(xa, y) = D(x, ya) + D(a, [x, y])$. Since $\widehat{Z}D(a, [x, y]) = 0$ it follows that $\widehat{M}'$ is a subbimodule of $\widehat{M}$.

We claim that $\widehat{M}_e + \widehat{M}_f \subseteq \widehat{M}'$. Indeed, $D(\widehat{M}_e, \widehat{M}_f) = 0$. Furthermore, $\widehat{Z}D(\widehat{M}_e, \widehat{M}_{ef}) \subseteq \widehat{Z} \cap \widehat{A}_{ef} = (0)$ and $\widehat{Z}D(\widehat{M}_e, \widehat{M}_e) \subseteq \widehat{Z} \cap \widehat{A}_e = (0)$. Since $\hat{M}_e + \widehat{M}_f$ generates $\widehat{M}$ we conclude that $\widehat{M}' = \widehat{M}$. Lemma is proved. □

Thus from Proposition 3.1 it follows that if $\mathcal{G}$ has capacity ≥ 3 then J is a loop superalgebra.

From now on in this chapter we will assume that $\mathcal{G} = F1 + V$ is a Jordan algebra of a symmetric nondegenerate bilinear form $<,>: V \times V \to F$. The case $dim_F V \leq 1$ has already been considered in chapter 3. Therefore we will assume that $dim_F V \geq 2$.

The gradation $\mathcal{G} = \mathcal{G}_0 + \mathcal{G}_1 + \cdots + \mathcal{G}_{n-1}$ is induced by a gradation $V = V_0 + V_1 + \cdots + V_{n-1}$. Indeed, $V = \mathcal{G}D(\mathcal{G}, \mathcal{G})$, that's why the subspace V of $\mathcal{G}$ is graded.

The center $Z = Z(A)$ is $F[t^{-n}, t^n]$.

An element $a \in \mathcal{L}(\mathcal{G})$ is refered to as an involution if $a^2 = 1$.

Let $0 < i < \frac{n}{2}$. The subspaces V_{n-i}, V_i are dual. Choose bases $w_{-i,1}, \cdots,$ $w_{-i,l}$ and $w_{i,1}, \cdots, w_{i,l}$ in V_{n-i} and V_i respectively such that
$< w_{-i,r}, w_{i,s} >= \delta_{r,s}$. Then the elements $\sqrt{\frac{1}{2}}(w_{-i,r}t^{-i} + w_{i,r}t^i)$ and
$\sqrt{\frac{-1}{2}}(w_{-i,r}t^{-i} - w_{i,r}t^i)$ are involutions.

Choose an orthonormal basis $w_{0,1}, w_{0,2}, \cdots$ in V_0.

If n is even and the dimension $\dim V_{n/2}$ is also even, choose a basis $w_{n/2,1}, \cdots,$ $w_{n/2,2s}$ in $V_{n/2}$ such that $< w_{n/2,2i-1}, w_{n/2,2i} >= 1$, $1 \leq i \leq s$ and all other products are equal to zero.
The elements $\sqrt{\frac{1}{2}}(w_{n/2,1}t^{-n/2} + w_{n/2,2}t^{n/2}), \cdots, \sqrt{\frac{1}{2}}(w_{n/2,2s-1}t^{-n/2} + w_{n/2,2s}t^{n/2})$ and $\sqrt{\frac{-1}{2}}(w_{n/2,1}t^{-n/2} - w_{n/2,2}t^{n/2}), \cdots, \sqrt{\frac{-1}{2}}(w_{n/2,2s-1}t^{-n/2} - w_{n/2,2s}t^{n/2})$ are involutions.

Let now $dim_F V_{n/2} = 2s + 1$. Again we choose a basis $\{w_{n/2,1}, \cdots, w_{n/2,2s},$ $u\}$ in $V_{n/2}$, where elements $w_{n/2,k}$'s have the same properties as above, the vector u is orthogonal to them and $u^2 = 1$.

We will call the system of involutions $\{v_1, \cdots, v_m\} = \{w_{0j}, \sqrt{\frac{1}{2}}(w_{-i,r}t^{-i} + w_{i,r}t^i), \sqrt{\frac{-1}{2}}(w_{-i,r}t^{-i} - w_{i,r}t^i), 0 < i < \frac{n}{2}, \sqrt{\frac{1}{2}}(w_{n/2,2k-1}t^{-n/2} + w_{n/2,2k}t^{n/2}),$ $\sqrt{\frac{-1}{2}}(w_{n/2,2s-1}t^{-n/2} - w_{n/2,2s}t^{n/2})\}$ a standard system of involutions of $\mathcal{L}(G)$. If $dim_F V_{n/2}$ is odd, then we will add $v_0 = ut^{n/2}$, though v_0 is not an involution, $v_0^2 = t^n$.

It is easy to see that for any $0 \leq i \neq j \leq m$ we have $v_i v_j = 0$ and therefore the operators $U(v_i), U(v_j)$ commute.

If $v \in A$ is an involution, the elements $e_1 = \frac{1}{2}(1 + v)$, $e_2 = \frac{1}{2}(1 - v)$ are orthogonal idempotents, $e_1 + e_2 = 1$. Let $M = M_1 + M_{1/2} + M_2$ be the Peirce decomposition of M with respect to e_1, e_2.

Remark. We have seen in the proof of Lemma 4.2 that $ZD(M_i, M_j) = 0$, unless $i = j = \frac{1}{2}$.

For an m-tuple $(i_1, \cdots, i_m)$, $i_k = \pm 1$, let $J(i_1, \cdots, i_m) = A(i_1, \cdots, i_m) + M(i_1, \cdots, i_m)$ denote the subspace of elements x such that $xU(v_k) = i_k x$. Since the involutive operators $U(v_k)$ commute, we have $J = \oplus J(i_1, \dots, i_m)$, $i_k = \pm 1$. Since $U(v_k)'s$ are automorphisms of the superalgebra J, it follows that $J(i_1, \dots, i_m) \cdot J(j_1, \dots, j_m) \subseteq J(i_1 j_1, \dots, i_m j_m)$.

The subspace $M(i_1, \dots, i_m)$ lies in the Peirce component $M_{1/2}$ with respect to $e_1 = \frac{1}{2}(1 + v_k)$, $e_2 = \frac{1}{2}(1 - v_k)$ if and only if $i_k = -1$. Otherwise it lies in $M_1 + M_2$.

In view of the remark above $ZD(M(\alpha), M(\beta))$ can be $\neq (0)$ only if $\alpha = \beta = (-1, -1, \dots, -1)$.

If $ZD(M, M) = (0)$ then by Proposition 3.1 J is a loop superalgebra. That's why from now on in this chapter, we will assume that

$$ZD(M, M) = ZD(M(-1, -1, \dots, -1),\ M(-1, -1, \dots, -1)) \neq (0).$$

As we have noticed above (see the proof of Lemma 3.20 , the graded subspace $ZD(M, M)$ contains all powers t^{nl}, $l \in \mathbb{Z}$, except, may be, one of them. This implies that for an arbitrary nonzero element $x \in M$, $x(ZD(M(-1, -1, \dots, -1), M(-1, -1, \dots, -1))) \neq (0)$. If $x \in M(\alpha)$, then the product above lies in $M(-1, -1, \dots, -1)A + ((M(-1, -1, \dots, -1))A)Z$ and in $M(\alpha)$ at he same time.

We have $A = Z + \sum_{i=1}^{m} Zv_i$. If $dim_F V_{n/2}$ is even (that is, there is no v_0) then $M(\alpha)$ can be $\neq 0$ only if $\alpha = (-1, -1, \dots, -1)$ or $\alpha = (1, \dots, 1, -1, \dots, 1)$. If there is $v_0 \in A(-1, -1, \dots, -1)$, then we can have also $M(1, 1, \dots, 1) \neq (0)$

CASE 1. $\dim V_{n/2}$ is even.

For $1 \leq i \neq j \leq m$, we have $M(1, \dots, 1, \underbrace{-1}_{i}, 1, \dots, 1)v_j \subseteq$
$M(-1, \dots, -1, \underbrace{1}_{i}, -1, \dots, -1, \underbrace{1}_{j}, -1, \dots, -1) = (0)$ except if $m = 3$ (in which case, for example, $M(-1, 1, 1)v_2 \subseteq M(1, 1, -1)$).

For an arbitrary element $x \in M(1, \dots, 1, \underbrace{-1}_{i}, 1, \dots, 1)$, we have $xU(v_j) = x$. Thus, for $m \neq 3$, $xv_j = 0$ and $xU(v_j) = x$ imply $x = xU(v_j) = x(2R(v_j)^2 - 1) = -x$, hence $x = 0$. Therefore $M = M(-1, -1, \dots, -1)$.

Now $M(Zv_1) \subseteq M(-1, 1, \dots, 1) = (0)$, which implies $D(M, Z) = D(M, (Zv_1)v_1) = (0)$ and therefore $ZD(M, M) = (0)$, which contradicts our assumption.

CASE 2. $\dim V_{n/2}$ is odd.

Now $M = M(-1, -1, \dots, -1) + \sum_i M(1, \dots, 1, \underbrace{-1}_{i}, 1, \dots, 1) + M(1, 1, \dots, 1)$.
As in the case 1, for an element $x \in M(1, \dots, 1, \underbrace{-1}_{i}, 1, \dots, 1)$ and $1 \leq j \leq m$, $i \neq j$, we have $xv_j \in M(-1, \dots, -1, \underbrace{1}_{i}, -1, \dots, -1, \underbrace{1}_{j}, -1, \dots, -1) = (0)$ except if $m \leq 3$.

Thus, again, if $m > 3$ then $M(1, \dots, 1, \underbrace{-1}_{i}, 1, \dots, 1) = (0)$ and
$M = M(-1, -1, \dots, -1) + M(1, 1, \dots, 1)$.

Now, $M(-1,-1,\ldots,-1)(Zv_1) \subseteq M(-1,1,\ldots,1) = (0)$, which implies that $D(M(-1,-1,\ldots,-1),Z) = D(M(-1,-1,\ldots,-1),(Zv_1)v_1) = (0)$ and therefore $ZD(M(-1,-1,\ldots,-1),M(-1,-1,\ldots,-1)) = ZD(M,M) = (0)$, which again contradicts our assumption.

The case $\dim V = 1$ was analized in Chapter 3. Thus the remaining cases are:

A) $\dim V_{n/2}$ odd and $m = 1, 2$ or 3, and

B) $\dim V_{n/2}$ even and $m = 3$.

The following lemma can be considered as an addition to Chapter 3.

LEMMA 4.3. *Let $J = A + M$ be a unital $\mathbb{Z}$-graded Jordan superalgebra such that $A = F[t^{-n}, t^n]$, $AD(M,M) \neq (0)$. Then J is graded simple.*

PROOF. Since A is a graded simple algebra, a proper ideal I of J should lie in M and $[I, M] = (0)$, $IA \subseteq I$.

For arbitrary elements $x, y \in M$, $a \in A$ we have $I(R([x,y])R(a) + R(xa)R(y) - R(ya)R(x) - R([xa,y]) - R(x)R(y)R(a) + R(a)R(y)R(x)) = (0)$ which implies $I(R([x,y])R(a) - R([xa,y]) = (0)$.

Exchanging x and y we get $I(R([y,x])R(a) - R([ya,x])) = (0)$, which implies $I(R([xa,y]) - R([x,ya]) = (0)$ or equivalently $I(AD(M,M)) = (0)$.

The subspace $AD(M,M)$ contains all homogeneous components of A except, may be, one (see the proof of Lemma 3.20). Hence $I = (0)$. Lemma is proved. □

<u>$dim_F V_{n/2}$ is odd, $m = 1$</u>

In this case $dim_F V = 2$, n is even and $A = Z + Zv_1 + Zv_0$, where $v_1 \in V_0$, $v_1^2 = 1$, $v_0 \in V_{n/2} \otimes t^{n/2}$, $v_0^2 = t^n$, $Z = F[t^{-n}, t^n]$, $v_1 v_0 = 0$.

Let $M = M(-1) + M(1)$ be the decomposition of M into a direct sum of eigenspaces with respect to $U(v_1)$.

There exists $x \in M(-1)$ such that $ZR(x)^2 \neq (0)$.

Then $M = M(ZR(x)^2) = xA + (xA)Z = (xZ)Z + (x(v_1Z))Z + (x(v_0Z))Z$.

The sum $J_1 = (Z + v_1Z) + M(-1)$ is a subsuperalgebra of J. From Proposition 3.31 it follows that $(yz')z'' = y(z'z'')$ for arbitrary elements $y \in M(-1)$, $z', z'' \in Z$. Let us show that $D(A,Z) = (0)$. Clearly, $D(Z,Z) = (0)$. For $i = 0, 1$ we have $Z = (v_iZ)^2$. Hence $D(v_jZ, Z) = D(v_jZ, (v_iZ)^2) = D((v_jZ)(v_iZ), v_iZ) = (0)$, $i \neq j$.

Since $x \in M(-1)$ we conclude that $M(R(z')R(z'') - R(z'z'')) = (0)$. let $U = U_M(v_0^{-1}, v_0)$. We have $U^2 = \frac{1}{2}Id_M + \frac{1}{2}U_M({v_0}^{-2}, {v_0}^2)$ and so in this case, $U_M(v_0^{-2}, v_0^2) = U_M(t^{-n}, t^n) = 2R(t^{-n})R(t^n) - Id_M = Id_M$. Hence, $U^2 = Id_M$.

Let $M = M_{-1} + M_1$ be the decomposition of M into a sum of eigenspaces with respect to U. Since the operators $U(v_1)$ and U commute, we have $M = \oplus_{i,j=\pm 1}(M(i) \cap M_j)$.

LEMMA 4.4. *There exists an element $y \in M(-1) \cap (M_{-1} \cup M_1)$ such that $t^n R(y)^2 \neq 0$.*

PROOF. Let us show that for an element $y = xf + (x(v_1 f_1))f_2 + (x(v_0 g_1))g_2$ we have $t^n R(y)^2 \neq 0$ if and only if $f \neq 0$. Indeed, we have $(x(v_0 g_1))g_2 \in M(1)$ and therefore $ZD(M, (x(v_0 g_1))g_2) = (0)$. Hence $t^n R(xf + (x(v_1 f_1))f_2 + (x(v_0 g_1))g_2)^2 = t^n R(xf + (x(v_1 f_1))f_2)^2$.

The element $xf + (x(v_1 f_1))f_2$ lies in $M(-1)$. From Corollary 3.38 applied to $J_1 = A(1) + M(-1)$ it follows that $ZD(M(-1), (x(v_1 Z))Z) = (0)$. Hence, $t^n R(xf + (x(v_1 f_1))f_2 + (x(v_0 g_1)g_2)^2 = t^n R(xf)^2$.

If an arbitrary element from $M(i) \cap M_j$ has a zero coefficient f at x, then $M = (x(v_1 Z))Z + (x(v_0 Z))Z$, which implies $ZD(M, M) = (0)$, the contradiction.

Hence, there exists $y \in M(i) \cap M_j$ such that $t^n R(y)^2 \neq 0$. Clearly, $i = -1$. Lemma is proved. □

In what follows, we assume that $M = xZ + (x(v_1 Z))Z + (x(v_0 Z))Z$, where $t^n R(x)^2 \neq 0$, $x \in M(-1)$, $xU = \xi x$, $\xi = \pm 1$.

We claim that $x(v_0 Z) \neq (0)$.

Otherwise $(v_0 Z)R_x^2 = (0)$ which implies $ZR(x)^2 = (v_0 Z)^2 R(x)^2 = (0)$, a contradiction.

Since $x(v_0 Z) \subseteq M(1)$ it follows that $(x(v_0 z))v_1 \neq 0$.

Comparing eigenvalues with respect to $U(v_1)$ we conclude that $(x(v_0 Z))v_1 \subseteq (x(v_0 Z))Z$.

Now compare eigenvalues with respect to U. The left-hand side has eigenvalue $-\xi$ whereas the right-hand side has eigenvalue ξ, a contradiction.

Hence this case is **impossible**.

$dim_F V_{n/2}$ is odd, $m = 2$

In this case $dim_F V = 3$, n is even and $A = Z + \sum_{i=0}^{2} Zv_i$, where $v_0^2 = t^n$, $v_0 \in V_{n/2} \otimes t^{n/2}$, $v_1, v_2 \in V_i \otimes t^i + V_{-i} \otimes t^{-i}$, $0 \leq i < n$.

(Notice that this included the cases $v_1, v_2 \in V_0$ and $v_1, v_2 \in V_{n/2} \otimes t^{n/2} + V_{n/2} \otimes t^{-n/2}$).

Let $M = \oplus M(i,j)$; $i, j = \pm 1$, be the decomposition of M into the sum of eigenspaces with respecto to $U(v_1), U(v_2)$.

Then $ZD(M(-1,-1), M(-1,-1)) \neq (0)$. Applying the results of chapter 3 to the subsuperalgebra $Z + M(-1,-1)$ we see that there exists a homogeneous element $x \in M(-1,-1)$ such that $t^n R(x)^2 \neq 0$ and therefore $M = (xZ)Z + \sum_{i=0}^{2} (x(v_i Z))Z$.

If $x(v_1 Z) = (0)$, then $D(x, Z) = D(x, (v_1 Z)(v_1 Z)) = (0)$ and therefore $ZR(x)^2 = (0)$. Hence, $x(v_1 Z) \neq (0)$. Since this subset belongs to the eigenvalue 1 with respect to $U(v_2)$ it follows that $(x(v_1 Z))v_2 \neq (0)$.

Comparing eigenvalues with respect to $U(v_1)$ and $U(v_2)$ we see that $(x(v_1 Z))v_2 \subseteq (x(v_0 Z))Z$.

Let $\deg(x) = \alpha$. Then $(x(v_1 Z))v_2 \subseteq \sum_{k \in \mathbb{Z}} M_{\alpha + kn} + \sum_{k \in \mathbb{Z}} M_{\alpha + kn \pm 2i}$ and $(x(v_0 Z))Z \subseteq \sum_{k \in \mathbb{Z}} M_{\alpha + n/2 + kn}$.

This is possible only if $i = \pm n/4$. Let $v_1 = \frac{1}{\sqrt{2}}(w_{n/4} \otimes t^{n/4} + w'_{-n/4} \otimes t^{-n/4})$, $v_2 = \frac{\sqrt{-1}}{\sqrt{2}}(w_{n/4} \otimes t^{n/4} - w'_{-n/4} \otimes t^{-n/4})$.

Let w be one of the elements $w_{n/4} \otimes t^{n/4}$ or $w'_{-n/4} \otimes t^{-n/4}$. Then $w^2 = 0$ and since w is a linear combination of v_1, v_2 we have $xw = 0$.

Hence, $xR(wZ)R(w) \subseteq \underbrace{xR(w)}_{0} R(wZ) + xD(wZ, w) = xD(Z, w^2) = (0)$.

This shows that summands from $M_{\alpha+kn\pm 2i}$ won't arise, since they are from $xR(wZ)R(w)$, where $w = w_{n/4} \otimes t^{n/4}$ or $w = w'_{-n/4} \otimes t^{-n/4}$.

So this case is **not possible**.

$dim_F V_{n/2}$ is odd, $m = 3$

In this case $A = Z + \sum_{i=0}^{3} Zv_i$ and as we have seen above $M = M(-1,-1,-1) + M(1,1,-1) + M(1,-1,1) + M(-1,1,1) + M(1,1,1)$. Choose an element $x \in M(-1,-1,-1)$ such that $ZR(x)^2 \neq (0)$. Then $M = (xZ)Z + + \sum_{i=0}^{3}(x(v_iZ))Z$.

We have $((x(v_0Z))Z)v_1 \subseteq M(1,-1,-1) = (0)$.

Since $(x(v_0Z))Z$ belongs to the eigenvalue 1 with respect to $U(v_1)$ it follows that $x(v_0Z) = (0)$.

This implies that $D(x, Z) = D(x, (Zv_0)^2) = (0)$, and hence, $ZR(x)^2 = (0)$, a contradiction.

This proves that this case is also **impossible**.

$dim_F V_{n/2}$ is even, $m = 3$

In this case $dim_F V = 3$ and $A = Z + \sum_{i=1}^{3} Zv_i$, with $v_3 \in V_0$, $v_1 = \frac{1}{\sqrt{2}}(w_i \otimes t^i + w_{-i} \otimes t^{-i})$, $v_2 = \frac{1}{\sqrt{-2}}(w_i \otimes t^i - w_{-i} \otimes t^{-i})$. (the case $V = V_0$ is also possible).

Let x be a homogeneous element from $M(-1,-1,-1)$ such that $t^nR(x)^2 \neq 0$.

Without loss of generality we will assume that $t^nR(x)^2 = t^n$ (in this case $deg(x) = 0$) or $t^nR(x)^2 = 1$ (in this case n is even and $deg(x) = \frac{-n}{2}$).

We have $M = (xZ)Z + (x(v_1Z))Z + (x(v_2Z))Z + (x(v_3Z))Z = M(-1,-1,-1) + M(-1,1,1) + M(1,-1,1) + M(1,1,-1)$.)

The subspace $J' = Z + (xZ)Z$ is a subsuperalgebra of J. By Lemma 4.1.3 J' is graded simple. In view of the results of chapter 3 we have to distinguish between two cases: (a) $x(t^n)^i \neq 0$ for any $i \in \mathbb{Z}$ and (b) $(xt^n)t^n = 0$.

Case (a)

By Proposition 3.30 $(xf)g = x(fg)$ for any $f, g \in Z$.

Let us show that $D(A, Z) = (0)$. Indeed, if $1 \leq i \neq j \leq 3$ then $Z = (v_jZ)^2$ and therefore $D(v_iZ, Z) = D(v_iZ, (v_jZ)^2) = (0)$.

This implies that $M(R(f)R(g) - R(fg)) = (0)$ for any $f, g \in Z$.

Let $' : Z \to Z$ be the derivation of Z such that $(t^n)' = 1$. We claim that

$$x(v_ig) = (x(v_it^n))g'$$

for an arbitrary element $g \in Z$.

Indeed, for $k \geq 1$ the equality $x(v_it^{nk}) = k(x(v_it^n))t^{n(k-1)}$ can be checked by the straightforward induction on k.

Similarly, $x(v_it^{-nk}) = k(x(v_it^{-n}))t^{-n(k-1)}$. By the Jordan identity $xR(v_it^{-n})R(t^n) = -xR(v_it^n)R(t^{-n})$, and therefore $xR(v_it^{-n}) = -xR(v_it^n)R(t^{-2n})$. Hence, $x(v_it^{-nk}) = -kx(v_it^n)t^{-n(k+1)} = (x(v_it^n))(t^{-nk})'$.

This implies that: $M = xZ + \sum_{i=1}^{3}(x(v_it^n))Z$ and $(xf)(v_ig) = (x(v_it^n))(fg')$ for arbitrary $f, g \in Z$.

Consider the element $(x(v_1t^n))v_2$. It is homogeneous of degree $deg(x) + n$. To prove this it is sufficient to check that $xR(wt^n)R(w) = 0$ where $w = w_i \otimes t^i$ or $w = w_{-i} \otimes t^{-i}$ if $i \neq 0$. Since w is a linear combination of v_1, v_2 and $x \in M(-1,-1,-1)$ we have $xw = 0$. By the Jordan identity $xR(wt^n)R(w) = \frac{1}{2}x(-R(w^2)R(t^n) + 2R(w)R(w)R(t^n) + R(w^2t^n)) = 0$.

Since $(x(v_1t^n))v_2 \in M(1,1,-1)$ it follows that $(x(v_1t^n))v_2 = (x(v_3t^n))f$, $f \in Z$. Comparing degrees we see that $f \in F$. Applying $R(v_2)$ to both sides of this equality we get

$$x(v_1t^n) = ((x(v_3t^n))v_2)f.$$

Now apply the derivation $D(v_3, v_1)$ to both sides of the last equality. We'll get

$$-x(v_3t^n) = ((x(v_1t^n))v_2)f = (x(v_3t^n))f^2.$$

Hence, $f^2 = -1$. Taking $-v_3$ instead of v_3 if necessary we can assume that $f = \sqrt{-1}$. Thus

$$(x(v_1t^n))v_2 = \sqrt{-1}x(v_3t^n).$$

Define the skew-symmetric cross product $V \times V \to V$ via $v_1 \times v_2 = v_3$, $v_1 \times v_3 = -v_2$, $v_2 \times v_3 = v_1$.

Then $(x(vt^n))w = \sqrt{-1}x((v \times w)t^n)$ for arbitrary elements $v, w \in V$.

We claim that $xR(v_if)R(v_ig) = 0$ for $1 \leq i \leq 3$; $f, g \in Z$.

Indeed, by the Jordan identity
$xR(v_if)R(v_ig) = x(-R(g)R((v_if)v_i) - R(v_i)R((v_if)g) + R(v_i)R(v_if)R(g) + R(g)R(v_if)R(v_i) + R((v_ig)(v_if)) = -xR(g)R(f) + xR(fg) = 0$.

Hence, $((x(v_it^n))Z)(v_iZ) = (0)$.

Now let $1 \leq i \neq j \leq 3$. For arbitrary elements $f, g \in Z$ we have $((x(v_it^n))f)(v_jg) = ((x(v_it^n))v_j)(fg) = \sqrt{-1}(x((v_i \times v_j)t^n))fg$ because $(x(v_it^n))f$ belongs to the eigenvalue 1 with respect to $U(v_j)$.

Finally, $((x(vt^n))f)(wg) = \sqrt{-1}(x((v \times w)t^n))(fg)$ for arbitrary elements $v, w \in V$.

We have determined the structure of an A-bimodule on M, that is, we know the multiplication table in A and the multiplication of elements of A by elements of M.

	g	v_jg
xf	xfg	$(x(v_jt^n))fg'$
$(x(v_it^n))f$	$(x(v_it^n))(fg)$	$\sqrt{-1}(x((v_i \times v_j)t^n))(fg)$

where $v_i \times v_i = 0$.

It remains to determine the bracket $[,]: M \times M \to A$.

Denote $D = R(x)^2$. Clearly, $D =' $ if $deg(x) = \frac{-n}{2}$ and $D =' R(t^n)$ if $deg(x) = 0$. It was checked in Chapter 3 that

$$[xf, xg] = D(f)g - fD(g)$$

for arbitrary elements $f, g \in Z$.

Notice that $ZD((x(v_i Z))Z, xZ) \subseteq Z \cap A(-1, -1, \dots, \underbrace{1}_{i}, \dots, -1) = (0)$.

Hence, for arbitrary elements $y' \in (x(v_i Z))Z$, $y'' \in xZ$, $a \in Z$ we have $[y'a, y''] = [y', y''a]$.

Let's verify that

$$[(x(v_i t^n))f, x] = \begin{cases} v_i f & \text{if deg(x)} = \frac{-n}{2} \\ v_i f t^n & \text{if deg(x)} = 0 \end{cases}$$

for an arbitrary element $f \in Z$.

Suppose at first that $f = t^{nk}$, $k \neq -1$. Then $f = h'$, where $h = \frac{1}{k+1} t^{n(k+1)}$ and therefore $(x(v_i t^n))f = x(v_i h)$. Now $[x(v_i h), x] = v_i D(h)$.

Let $k = -1$. We have

$$[(x(v_i t^n))t^{-n}, x]t^{-n} = x(v_i t^n)R(t^{-n})R(x)R(t^{-n}) =$$

$$\frac{1}{2} x(v_i t^n)(-R(xt^{-2n}) + 2R(t^{-n}R(xt^{-n}) + R(x)R(t^{-2n})) =$$

$$\frac{-1}{2}[x(v_i t^n), xt^{-2n}] + [x(v_i t^n)t^{-n}, xt^{-n}] + \frac{1}{2}[x(v_i t^n), x]t^{-2n} =$$

$$\frac{-1}{2}[(x(v_i t^n))t^{-2n}, x] + [(x(v_i t^n))t^{-2n}, x] + \frac{1}{2}[x(v_i t^n), x]t^{-2n} = \begin{cases} v_i t^{-2n} \\ v_i t^{-n} \end{cases}$$

depending of deg(x). This implies that

$$[(x(v_i t^n))t^{-n}, x] = \begin{cases} v_i t^{-n} & \text{if deg(x)} = \frac{-n}{2} \\ v_i & \text{if deg(x)} = 0 \end{cases}$$

Now for an arbitrary element $g \in Z$ we have

$$[(x(v_i t^n))f, xg] = [(x(v_i t^n))(fg), x] = \begin{cases} v_i fg & \text{if deg(x)} = \frac{-n}{2} \\ v_i fg t^n & \text{if deg(x)} = 0 \end{cases}$$

Notice that

$$D((x(v_i Z))Z, Z) = D((x(v_i Z))Z, (v_i Z)^2) = (0)$$

because $((x(v_i Z))Z)(v_i Z) = (0)$.

This implies that for $f, g \in Z$ we have

$$[(x(v_i t^n))f, (x(v_i t^n))g] = [x(v_i t^n), x(v_i t^n)]fg = 0.$$

Similarly, for $1 \leq i \neq j \leq 3$ we have

$$[(x(v_i t^n))f, (x(v_j t^n))g] = [x(v_i t^n), x(v_j t^n)]fg = 0.$$

Indeed,

$$[x(v_it^n), x(v_jt^n)] = (v_it^n)D(x, x(v_jt^n)) - [(x(v_jt^n))(v_it^n), x] =$$
$$v_iD(x, x(v_jt^n))t^n - [\sqrt{-1}(x((v_j \times v_i)t^n))t^n, x] =$$
$$\sqrt{-1}[x((v_j \times v_i)t^n), x]t^n - \sqrt{-1}[x((v_j \times v_i)t^n)t^n, x] = 0.$$

We have the following tables:

(T.4.1) deg(x) = 0

	xg	$(x(v_jt^n))g$
xf	$(f'g - fg')t^n$	$-v_jfgt^n$
$(x(v_it^n))f$	v_ifgt^n	0

(T.4.2) deg(x) = -n/2

	xg	$(x(v_jt^n))g$
xf	$f'g - fg'$	$-v_jfg$
$(x(v_it^n))f$	v_ifg	0

The tables T.4.1 and T.4.2 define Jordan superalgebra structures on $J = A+M$, (these superalgebras are embeddable into Kantor doubles, see Chapter 6) The Tits-Kantor-Koecher constructions of these Jordan superalgebras are the conformal Lie superalgebras CK(6) (of Neveu-Schwarz and Ramond types respectivley) introduced by Cheng and Kac in [CK].

Case (b)

Now let us assume that $(xt^n)t^n = 0$ and therefore $(Mt^n)t^n = (0)$.

Clearly, $[x(v_1t^n), x(v_1t^n)] = 0$. On the other hand, $[x(v_1t^n), x(v_1t^n)] = (v_1t^n)R(x)R(x(v_1t^n)) = (v_1t^n)D(x, x(v_1t^n)) - (v_1t^n)R(x(v_1t^n))R(x)$.

We have $t^nD(x, x(v_1t^n)) \in Z \cap A(1, -1, -1) = (0)$ and $v_1D(x, x(v_1t^n)) = 0$ because $xv_1 = 0$ and $(x(v_1t^n))v_1 = (xv_1)(v_1t^n) = 0$.

Hence, $[(x(v_1t^n))(v_1t^n), x] = 0$.

By the Jordan identity

$xR(v_1t^n)R(v_1t^n) = x(-R(v_1)R(v_1t^{2n}) - R(t^n)R((v_1t^n)v_1) + 2R(v_1)R(v_1t^n)R(t^n) + R(t^{2n})) = xt^{2n}$.

We showed that $[xt^{2n}, x] = 0$, the contradiction. Hence, this case is impossible.

CHAPTER 5

J is a finite dimensional Jordan Superalgebra or a Jordan Superalgebra of a Superform

In this chapter we will consider all cases when J is finite dimensional or a Jordan superalgebra of a superform.

5.1. A is finite dimensional

PROPOSITION 5.1. Let $J = A + M$ be a graded simple Jordan superalgebra and $dim_F A < \infty$. Then $dim_F J < \infty$ or J is a Jordan superalgebra of a superform.

First we will notice that J is locally finite dimensional, that is, every finitely generated subalgebra of J is finite dimensional. Indeed, let $X = \{x_1, \dots, x_k\}$ be a finite subset of M. Then the subbimodule $\sum_{i=1}^{k} x_i R < A >$ of M is finite dimensional (see [J3]) and $A + \sum_{i=1}^{n} x_i R < A >$ is a finite dimensional subalgebra of J.

The multiplication algebra $R < J >$ is $\mathbb{Z}/2\mathbb{Z}$-graded, $R < J >=$ $R < J >_{\bar{0}} + R < J >_{\bar{1}}$, where $R < J >_{\bar{0}}$ consists of operators that map $A \to A$, $M \to M$, while elements from $R < J >_{\bar{1}}$ map $A \to M$, $M \to A$.

Moreover, the $\mathbb{Z}$-gradation on J induces a $\mathbb{Z}$-gradation on $R < J >$, $R < J >_{\bar{0}} = \sum_{i \in \mathbb{Z}} R < J >_{\bar{0},i}$, $R < J >_{\bar{1}} = \sum_{i \in \mathbb{Z}} R < J >_{\bar{1},i}$.

LEMMA 5.2. *If $x_1, \dots, x_k$ are linearly independent elements from a homogeneous component M_n (resp. A_n), $n \in \mathbb{Z}$, then there exist operators $\omega_1, \dots, \omega_k \in R < J >_{\bar{1},-n}$ (resp. $R < J >_{\bar{0},-n}$) such that $x_i \omega_j = \delta_{ij}$.*

PROOF. Since the superalgebra J is graded simple it follows that M_n (resp. A_n) is an irreducible module over $R < J >_{\bar{0}0}$.
By Density Theorem (see [J1]) there exist homogeneous operators $\omega'_1, \dots, \omega'_k \in R < J >_{\bar{0}0}$ such that $x_i \omega'_j = 0$ for $i \neq j$ and $x_i \omega'_i \neq 0$. Again, since J is graded simple it follows that there exist homogeneous operators $\omega''_1, \dots, \omega''_k \in R < J >$ such that $x_i \omega'_i \omega''_i = 1$. Let $\omega_i = \omega'_i \omega''_i$. It is easy to see that $\omega_i \in R < J >_{\bar{1},-n}$ (resp. $R < J >_{\bar{0},-n}$). Lemma is proved.

□

LEMMA 5.3. *A graded simple finite dimensional Jordan superalgebra is simple.*

PROOF. Let S be a graded simple Jordan superalgebra, $dim_F S < \infty$. Since the solvable radical $Rad(S)$ is $Aut(S)$-invariant it follows that $Rad(S)$ is graded. Hence $Rad(S) = (0)$.

Let I be a proper (not graded) ideal of S. Consider the descending chain of ideals $I^{(1)} = I$, $I^{(i+1)} = (I^{(i)})^3$. There exists $k \geq 1$ such that $I^{(i+1)} = I^{(i)}$. Since $Rad(S) = (0)$ it follows that $I^{(i)} \neq (0)$. Thus, without loss of generality we can assume that $I = I^3$. This implies that I is $Der(S)$-invariant, hence graded. This contradicts our assumption that S is graded simple. Lemma is proved. □

From the classification of simple finite dimensional Jordan algebras it follows that there is a function $f(m)$ such that if $S = S_{\bar{0}} + S_{\bar{1}}$ is a simple finite dimensional Jordan superalgebra and $dim_F S_{\bar{0}} \leq m$ then $dim_F S \leq f(m)$ or S is a Jordan superalgebra of a superform.

Suppose that M is infinite dimensional and let $k > f(dim_F A)$. Choose k homogeneous linearly independent elements $x_1, \dots, x_k \in M$.

Choose arbitrary linearly independent homogeneous elements from $a_1, \dots, a_q$ from A. Suppose that one of the elements $a_1, \dots, a_q, x_1, \dots, x_k$ lies in $J_{i\bar{\sigma}}$. Choose all elements from $a_1, \dots, a_q, x_1, \dots, x_k$ lying in $J_{i\bar{\sigma}}$ and for these elements find operators guaranteered by Lemma 5.1. We will get $k + q$ homogeneous operators $\omega_1, \dots, \omega_{k+q} \in R < J >$.

Let J' be a finite dimensional subalgebra of J containing $1, a_1, \dots, a_q, x_1, \dots, x_k$ and all elements involved in $\omega_1, \dots, \omega_{k+q}$.

Since $1 \in J'$ the superalgebra J' contains a maximal graded ideal $P(J')$, so $J'/P(J')$ is graded simple. We claim that the images of the elements $x_1, \dots x_k$ in $J'/P(J')$ are linearly independent. Indeed, suppose that $\alpha_{i_1} x_{i_1} + \cdots + \alpha_{i_r} x_{i_r} \in P(J')$. Since the ideal $P(J')$ is graded we can assume that all elements $x_{i_1}, \dots, x_{i_r}$ have the same degree. By the choice of J' the operators $\omega_{i_1}, \dots, \omega_{i_r}$ corresponding to $x_{i_1}, \dots, x_{i_r}$ via Lemma 5.1.1 lie in $R < J' >$. Hence $\alpha_{i_t} 1 = (\alpha_{i_1} x_{i_1} + \cdots + \alpha_{i_r} x_{i_r}) \omega_{i_t} \in P(J')$, hence $\alpha_{i_t} = 0$.

Similarly the images of the elements $a_1, \dots, a_q$ in $J'/P(J')$ are linearly independent.

By Lemma 5.2 the superalgebra $S = J'/P(J')$ is simple. Since $dim_F S_{\bar{1}} \geq k > f(dim_F S_{\bar{0}})$ it follows that $J'/P(J')$ is a superalgebra of a bilinear nondegenerate superform.

We claim that the algebra A is quadratic.

Indeed, let $a \in A$ and suppose that the elements $1, a, a^2$ are linearly independent. Again choose k linearly independent elements $x_1, \dots, x_k \in M$, $k > f(dim_F A)$. As we have seen above there exists a finite dimensional subalgebra $J' \leq J$ and a maximal ideal $P(J') \triangleleft J'$ such that

(1) the images of $1, a, a^2, x_1, \dots, x_k$ in $J'/P(J')$ are linearly independent,

(2) $J'/P(J')$ is a superalgebra of a bilinear nondegenerate superform.

The assertions (1),(2) contradict each other because the even part of a superalgebra of a superform is quadratic.

An arbitrary element $a \in A - F1$ satisfies exactly one quadratic equation $a^2 - 2t(a)a + n(a)1 = 0$. For $\alpha \in F1$, let $t(\alpha) = \alpha$. Then $t : A \to F$ is a linear mapping (see [ZSSS]) and $A = F1 + V$,where $V = \{a \in A | t(a) = 0\}$.

Our next claim is that $V.M = (0)$.

Indeed, let $v \in V$, $x \in M$ and $xv \neq 0$. Again we can find a finite dimensional subalgebra J' of J and a maximal ideal $P(J')$ such that $A \subseteq J'$, $x \in J'$, $xv \notin P(J')$, v and 1 are linearly independent modulo $P(J')$ and $J'/P(J')$ is a Jordan superalgebra of a superform, what brings us a contradiction.

To finish the proof that J is a superalgebra of a superform we need only to check that $[M, M] \subseteq F1$. Suppose that $x, y \in M$ and the elements $[x, y]$ and 1 are linearly independent. Again we can find a finite dimensional superalgebra J' and a maximal ideal $P(J')$ such that $A \subseteq J'$, $x, y \in J'$, $J'/P(J')$ is a Jordan superalgebra of a superform and the elements $[x, y]$ and 1 are linearly independent modulo J'. This brings us a contradiction. Proposition 5.1.1 is proved.

Remark. This proof shows that if $J = A + M$ is a simple (not necessarily graded) Jordan superalgebra and $dim_F A < \infty$ then $dim_F J < \infty$ or J is a Jordan superalgebra of a superform.

5.2. A/I finite dimensional, $I \neq (0)$

PROPOSITION 5.4. Under the assumptions above the superalgebra J is finite dimensional.

We showed in Chapter 1 that if $dim_F(A/I) < \infty$ then $\bar{A} = A/I$ is a simple Jordan algebra of a nondegenerate symmetric bilinear form. Let $\bar{A} = F1 + \bar{V}$, $<,>: \bar{V} \times \bar{V} \to F$ is the bilinear form and let V be the preimage of $\bar{V}$ under the homomorphism $A \to \bar{A}$, $A = V + F1$.

LEMMA 5.5. *For arbitrary elements $x, y \in M$, $a \in A$ we have*

$$M(xD(y, a)) \subseteq xR < A > +yR < A > .$$

PROOF. Choose an arbitrary element $m \in M$. Then $m(xD(y, a)) = [m, x]D(y, a) + (mD(y, a))x = (y[m, x])a - y([m, x]a) + x(mD(y, a)) \in xR < A > +yR < A >$. Lemma is proved. □

LEMMA 5.6. $dim_F M/(MI)A < \infty$.

PROOF. If $[(MI)A, M] \subseteq I$, then $id_J(I) \cap A = I$. Since the superalgebra J is simple we conclude that $[(MI)A, M]$ is not contained in I. By Lemma 1.28 $t([(mI)A, M]) = (0)$. This implies that $[(MI)A, M] \subseteq V$.

Since $\bar{V}$ is an irreducibe $D(\bar{A}, \bar{A})$-module, $\overline{[(MI)A, M]}$ is $D(\bar{A}, \bar{A})$-invariant and $[MI, M] \subseteq I$, it follows that $\overline{[(MI)A, M]} = \overline{(MI)D(A, M)} = \bar{V}$.

Since $dim_F \bar{V} < \infty$ we can choose $x_i, y_i \in M$, $u_i \in I$, $a_i \in A$ such that the elements $(x_i u_i)D(a_i, y_i)$ span V modulo I. Then by Lemma 5.5, $M.V \subseteq (MI)A + \sum_i x_i R < A > + \sum_i y_i R < A >$.

Now choose elements $x, y \in M$, $u \in I, a \in A$ such that $\overline{[(xu)a, y]} = \bar{v} \neq 0$. Then there exists $w \in V$ such that $< \bar{v}, \bar{w} >= 1$. The element $((xu)a)D(y,w)$ has trace 1. Hence, $1 \in ((xu)a)D(y,w) + V$. Now we have $M = M1 \subseteq xR < A > +yR < A > + \sum_i x_i R < A > + \sum_i y_i R < A > +(MI)A$.

We proved that the A-bimodule $M/(MI)A$ is finitely generated. In fact, $M/(MI)A$ is an $\bar{A}$-module. Since $dim_F \bar{A} < \infty$ it follows that $dim_F M/(MI)A < \infty$. Lemma is proved. □

Let $A = I + Fa_1 + \cdots + Fa_t$, $M = (MI)A + Fx_1 + \cdots + Fx_s$.

Let us consider the set W of all products of right multiplications $R(a_i), R(x_j)$ such that:

(1) There are no three operators $R(a_i)$, $R(a_j)$, $R(a_k)$ standing together,

(2) There are no three operators $R(x_i)$, $R(x_j)$, $R(x_k)$ standing together and

(3) Every $R(a_i)$ occurs ≤ 2 times.

Clearly W is finite, because the length of any operator in W is $\leq 2t + (2t+1)2 = 6t + 2$.

For any $w \in W$ the kernel of $\phi_w : a \in I \to t(aw)$ has codimension ≤ 1. Let $\tilde{I} = \cap_{w \in W} Ker \phi_w$, $[I : \tilde{I}] < \infty$. Suppose that $\tilde{I} \neq (0)$, $0 \neq a \in \tilde{I}$.

Then there exists a multiplication operator $w \in R < J >$ such that $t(aw) \neq 0$. Let w have the minimal length among all operators with this property.

Then we can assume that all even elements that appear in w lie in $\{a_1, \ldots, a_t\}$. Indeed, if at least one even element lies in I, then (modulo operators of shorter length) we can move it to the right end, in which case the result lies in $(AI)A \subseteq I$ or in $[MI, M] \subseteq I$.

Suppose that one of the odd elements in w lies in $(MI)A$. Again without loss of generality we will assume that this element is either at the right end of w or just to the left of the right end. If it is at the right end, then $aw \in [M, (MI)A]$ and $t(aw) = 0$ by Lemma 1.28.

Let $w = R(y_1) \cdots R(y_r)$, $y_i \in A \cup M$ and suppose that $y_{r-1} \in (MI)A$. Then $y_r \in A$. Indeed, let $y_r \in M$.

We have $aw = -aR(y_1) \cdots R(y_r)R(y_{r-1}) + aR(y_1) \cdots R(y_{r-2})D(y_{r-1}, y_r)$.

The first summand has zero trace by Lemma 1.28. The second summand lies in $AD(M, M) \subseteq V$ and thus also has zero trace.

We claim that $y_{r-2} \in A$. Indeed, let $y_{r-2} \in M$. From the Jordan identity it follows that $aw = R(y_1) \cdots R(y_r)R(y_{r-1})R(y_{r-2}) + \sum_i w_i$, where w_i are operators of length $< r$. By minimality of r $tr(aw_i) = 0$.

Now, $aR(y_1) \cdots R(y_r)R(y_{r-1})R(y_{r-2}) \in [(MI)A, M]$ and thus $t(aR(y_1) \cdots R(y_r)R(y_{r-1})R(y_{r-2})) = 0$.

The operator w is even, whereas the operator $R(y_{r-2})R(y_{r-1})R(y_r)$ is odd. Hence $r \geq 4$.

We claim that $y_{r-3} \in M$. Indeed, let $y_{r-3} \in A$. Then $w \in \cdots R(A)R(A)R(M)R(A)$. Using the Jordan identity we can get three even multiplications at the right end. By the identity (D) (see the Introduction) $w \in w'D(A,A) + \sum_i w_i$, where w' is an operator of length $r-2$ and w_i are operators of length $< r$. This implies $t(aw) = 0$.

Hence $aR(y_1)\cdots R(y_{r-4}) \in A$. From the minimality of r it follows that $t((aR(y_1)\cdots R(y_{r-4}))A) = (0)$. Since the form $t(ab)$ is nondegenerate on A/I we conclude that $aR(y_1)\cdots R(y_{r-4}) \in I$. Now $aw \in [(MI)A, (MI)A]A \subseteq I$ by Lemma 1.41.

We proved that the element y_{r-1} can not lie in $(MI)A$. Hence we can assume that $y_1, \ldots, y_r \in \{a_1, \ldots, a_t, x_1, \ldots, x_s\}$.

If w contains three right multiplications of the same parity standing together then by the Jordan identity we can move them to the right end. By the identity (D), $w \in AD(M,M) + AD(A,A) + \sum_i w_i$, where w_i are operators of length $< r$. Hence $t(aw) = 0$, the contradiction.

If $R(a_i)$ occurs in w three times, then again by the Jordan identity w is a linear combination of operators of length $< r$.

Hence, $w \in W$. Now $t(aw) \neq 0$ contradicts the inclusion $a \in \tilde{I}$.

We proved that $\tilde{I} = (0)$.

Hence $dim_F I < \infty$ and therefore $dim_F A < \infty$. By Proposition 5.1.1 $dim_F J < \infty$ or J is a Jordan superalgebra of a superform. In the latter case, though, $I = (0)$. Proposition 5.2.1 is proved.

5.3. A is a Jordan algebra of a bilinear form

The aim of this section is to prove the following proposition

PROPOSITION 5.7. Let $J = A + M$ be a graded simple Jordan superalgebra such that $A = F1 + V$ is a Jordan algebra of a symmetric bilinear nondegenerate form $<,>$ on a vector space V. Then $dim_F J < \infty$ or J is a Jordan superalgebra of a superform.

PROOF. If $dim_F V < \infty$ then the proposition follows from Proposition 5.1. So suppose that V is infinite dimensional.

Let $n \geq 5$ be an odd number and choose n elements $v_1, \ldots, v_n \in V$ such that $< v_i, v_j >= \delta_{ij}$. The operators $U(v_i) : J \to J$ are involutive automorphisms. Let G be the group generated by $U(v_1), \ldots, U(v_n)$. Since they commute, the group G is abelian, so both, A and M decompose into direct sums $A = \oplus A(\epsilon_1, \ldots, \epsilon_n)$, $M = \oplus M(\epsilon_1, \ldots, \epsilon_n)$ and $J(\epsilon_1, \ldots, \epsilon_n) = A(\epsilon_1, \ldots, \epsilon_n) + M(\epsilon_1, \ldots, \epsilon_n) = \{x \in J | xU(v_i) = \epsilon_i x, 1 \leq i \leq n\}$.

Notice that for any $(\epsilon_1, \ldots, \epsilon_n)$, $(\mu_1, \ldots, \mu_n) \in \{\pm 1\}^n = \mathcal{A}$ we have $J(\epsilon_1, \ldots, \epsilon_n)J(\mu_1, \ldots, \mu_n) \subseteq J(\epsilon_1\mu_1, \ldots, \epsilon_n\mu_n)$.

Let us consider $\mathcal{A}_1 = \{(\epsilon_1, \ldots, \epsilon_n) \in \mathcal{A} | \prod_{i=1}^n \epsilon_i = 1\}$ and $\mathcal{A}_{-1} = \{(\epsilon_1, \ldots, \epsilon_n) \in \mathcal{A} | \prod_{i=1}^n \epsilon_i = -1\}$.

It is clear that $\mathcal{A}_1\mathcal{A}_1 \subseteq \mathcal{A}_1$, $\mathcal{A}_1\mathcal{A}_{-1} \subseteq \mathcal{A}_{-1}$, $\mathcal{A}_{-1}\mathcal{A}_{-1} \subseteq \mathcal{A}_1$.

Consider the subalgebra $B = F1 + \sum_{i=1}^n Fv_i$ of A. Let

$$M' = \oplus\{M(\alpha_1, \ldots, \alpha_n) | (\alpha_1, \ldots, \alpha_n) \in \mathcal{A}_1\},$$
$$M'' = \oplus\{M(\alpha_1, \ldots, \alpha_n) | (\alpha_1, \ldots, \alpha_n) \in \mathcal{A}_{-1}\}.$$

Then both M' and M'' are B-submodules and $B + M'$ and $B + M''$ are sub-superalgebras of J.

Indeed, for arbitrary two root vectors x, y from M' (resp. M'') their product $[x, y] \in \oplus A(\epsilon_1, \ldots, \epsilon_n)$, where $(\epsilon_1, \ldots, \epsilon_n) \in \mathcal{A}_1$.

Let $1 \leq i \leq n$. Then $v_i \in A(-1, \ldots, \underbrace{1}_{i}, \ldots, -1)$ and $(-1, \ldots, 1, \ldots, -1) \in \mathcal{A}_1$; $1 \in A(1, 1, \ldots, 1)$. If $v \in (v_1, \ldots, v_n)^\perp$ then $v \in A(-1, -1, \ldots, -1)$ and $(-1, \ldots, -1) \in \mathcal{A}_{-1}$.

So $\oplus\{A(\epsilon_1, \ldots, \epsilon_n) | (\epsilon_1, \ldots, \epsilon_n) \in \mathcal{A}_1\} = F1 + \sum_{i=1}^n Fv_i = B$.
We have shown that $[M', M'] \subseteq B$ and $[M'', M''] \subseteq B$.

Let $\tilde{M}' = \{x \in M' | [x \cdot B, M'] = (0)\}$. It is easy to see that $\tilde{M}'$ is a B-subbimodule of M' and that the superalgebra $B + M'/\tilde{M}'$ is simple. From the remark after the Proposition 5.1 it follows that $B + M'/\tilde{M}'$ is either finite dimensional or a Jordan algebra of a superform. Looking at the classification list of all finite dimensional Jordan superalgebras we see that only a Jordan superalgebra of a superform has its even part isomorphic to a Jordan algebra of a symmetric bilinear nondegenerated form in a vector space of dimension ≥ 5. Hence, $B + M'/\tilde{M}'$ is a Jordan superalgebra of a superform. Hence, $[M'/\tilde{M}', M'/\tilde{M}'] \subseteq F1$, which implies that $[M', M'] \subseteq F1$. Similarly, $[M'', M''] \subseteq F1$.

$[M', M''] \subseteq \oplus\{A(\epsilon_1, \ldots, \epsilon_n) | (\epsilon_1, \ldots, \epsilon_n) \in \mathcal{A}_{-1}\} = (v_1, \ldots, v_n)^\perp = \{v \in V | < v, v_i >= 0, 1 \leq i \leq n\} = U$. This implies that $[M, M] = [M', M'] + [M', M''] + [M'', M''] \subseteq F1 + U$. We can do the same with any finite set of vectors. Notice that when we consider another set of vectors $\{w_1, \ldots, w_m\}$ the submodules M' and M'' will change, but M does not change. So finally we get that $[M, M] \subseteq F1$.

Since $B + M'/\tilde{M}'$ is a superalgebra of a superform, for any i, $1 \leq i \leq n$, we have $M' \cdot v_i \subseteq \tilde{M}'$. Hence, $[M' \cdot v_i, M] = [M' \cdot v_i, M'] + [M' \cdot v_i, M''] = (0)$. Similarly, $[M'' \cdot v_i, M] = (0)$. Hence, $[M \cdot V, M] = (0)$. Now, $M \cdot V$ is an ideal of J and therefore $M \cdot V = (0)$. Proposition is proved. □

CHAPTER 6

The Main Case

In this chapter we will consider the main case of Proposition 1.26, that is, the case when $I \neq (0)$; $A/I = \mathcal{L}(G)$ the loop algebra associated to finite dimensional Jordan algebra of a symmetric nondegenerate bilinear form. In this case $I = M(A)$.

We will divide the chapter in two sections. In the first section we will prove that the algebra A splits, that is, there exists a graded subalgebra B of A, $B \simeq \mathcal{L}(G)$ such that $A = I + B$. In the second section we will determine the structure of J.

6.1. Splitting Theorem

At first we will prove an extension of Lemma 1.41

LEMMA 6.1. *Let $\tilde{Z}$ denotes the preimage of the center $Z(A/I)$ under the natural homomorphism $A \longrightarrow A/I \simeq \mathcal{L}(G)$, then $[(MI)A, MD(\tilde{Z}, A)] \subseteq I$.*

PROOF. Fix an element $c \in \tilde{Z}$ and consider the mapping $f : M \times M \times I \times A \times A \rightarrow A/I$, $f(x, y, u, a, b) = [(xu)a, yD(c, b)] + I/I$, where $x, y \in M$; $u \in I$; $a, b \in A$.

We claim that for arbitrary elements x, y, u, a, b
$f(x, y, u, a, b) = -f(x, y, u, b, a)$ and
$f(x, y, u, a^2, b) = 2f(xa, y, u, a, b) = 2f(x, ya, u, a, b) = 2f(x, y, ua, a, b) = 2f(x, y, u, a, ab)$.

Indeed, suppose that $a = b$. We have $[(xu)a, yD(c, a)] = [(xu)aD(a, c), y] \bmod I$. By the Jordan identity for arbitrary elements $a', a'' \in A$ we have $R(u)R(a')R(a'') = R(u)R(a'a'') \bmod R < A > R(I)$. Hence by Lemmas 1.27, 1.34 $[(xu)R(a)R(a)R(c), y] = [(xu)R(a^2)R(c), y] = [(xu)Ra^2c), y] \bmod I$ and $[(xu)R(a)R(c)R(a), y] = \frac{1}{2}[(xu)(-R(a^2c) + 2R(ac)R(a) + R(a^2)R(c), y] = [(xu)R(\frac{-1}{2}a^2c + (ac)a + \frac{1}{2}a^2c), y] \bmod I$ and therefore $f(x, y, u, a, a) = 0$.

Applying the Jordan identity to $R(u)R(a^2)$ we see that $f(x, y, u, a^2, b) = 2f(x, y, ua, a, b)$.

This implies $f(x, y, u, a^2, a) = 2f(x, y, ua, a, a) = 0$ and therefore $f(x, y, u, a^2, b) = 2f(x, y, u, a, ab)$. Again by the Jordan identity $xR(u)R(a^2) = uR(x)R(a^2) = u(-2R(a)R(xa) + 2R(xa)R(a) + R(a^2)R(x)) = 2(xa)R(u)R(a) \bmod MI$. This implies that $f(x, y, u, a^2, b) = 2f(xa, y, u, a, b)$.

And finally,

$$R(a^2)R(yD(c, b)) = -2R(yD(c, b)a)R(a) + R(yD(c, b))R(a^2) + 2R(a)R(yD(c, b)a),$$

where $yD(c,b)a = (ya)D(c,b) - yD(aD(c,b))$, $aD(c,b) \in I$. Hence,

$$f(x,y,u,a^2,b) = 2f(x,ya,u,a,b).$$

As we have mentioned in the proof of Lemma 1.34, unless $A = \tilde{Z}$ the algebra A contains two (not necessarily homogeneous) orthogonal idempotents e_1, e_2; $e_1 + e_2 = 1$. If $A = \tilde{Z}$ then $[(MI)A, M] \subseteq I$ by Lemma 1.34. Let $e_1, e_2 \in A$ and let $A = A_{11} + A_{12} + A_{22}$ be the corresponding Peirce decompositions.

In the proof of Lemma 1.34 we showed that the properties of the function f that we have established above imply $f(M,M,I,A_{11},A_{11}) = f(M,M,I,A_{22},A_{22}) = (0)$.

Let us show that $f(x,y,u,a,e_i) = 0$ for any $x, y \in M$; $u \in I$, $a \in A, i = 1,2$. Indeed, since $f(x,y,u,a,e_i) = 2f(x,ye_i,u,a,e_i)$ we can assume that $y \in M_{12}$. Let $c = c_{11} + c_{12} + c_{22}$ be the Peirce decomposition of c. We have $yD(c_{11} + c_{22}, e_i) = 0$. From $c_{12} \in I$ it follows that $yD(c_{12}, e_i) \in MD(I,A)$. Hence

$$[xR(u)R(a), yD(c,e_i)] = [xR(u)R(a), yD(c_{12},e_i)] \in [(MI)A, (MI)A] \subseteq I$$

by Lemma 1.41.

Now $f(M,M,I,A_{12},A_{11}) = f(M,M,I,A_{12}e_2,A_{11}) \subseteq$
$f(M,M,I,A_{12},A_{11}e_2) + f(M,M,I,e_2,A_{12}A_{11}) = (0)$ and similarly
$f(M_{ii},M,I,A_{12},A) = f(M,M,I_{ii},A_{12},A) = (0)$.

Hence, $f(M,M,I,A,A) = f(M_{12},M_{12},I_{12},A_{12},A_{12})$.

This implies that $[(MI)A, MD(\tilde{Z},A)] \subseteq A_{12} + I$.

The subspace $[(MI)A, MD(\tilde{Z},A)] + I/I$ of $\mathcal{L}(G)$ is invariant under all inner derivations of $\mathcal{L}(G)$. The only nonzero invariant subspaces are $Z(\mathcal{L}(G))$, $\sum_{i=j \ mod \ n} V_i \otimes t^j$, $\mathcal{L}(G)$ and none of them is contained in the Peirce component $(\mathcal{L}(G))_{12}$. Lemma is proved. □

Let $a \in I$ be a nonzero element. Then there exists a multiplication operator $w \in R < J >$ such that $aw \in A - I$. Indeed, otherwise the ideal $id_J(a)$ lies in $I + M$, which contradicts the simplicity of J.

DEFINITION 6.2. An operator $w = R(b_1) \cdots R(b_r)$ is a rescuing operator of the element $a \in I$ if

(1) $b_i \in A \cup M$, $1 \leq i \leq r$,

(2) $aW \in A - I$,

(3) for every subset $\{c_1, \ldots, c_s\} \subseteq A \cup M$ with $s < r$ either $aR(c_1) \cdots R(c_s) \in M$ or $aR(c_1) \cdots R(c_s) \in I$.

LEMMA 6.3. *A rescuing operator w belongs to one of the two following types:*
(i) $w = R(x_0)R(a_1)R(x_1) \cdots R(a_l)R(x_l)$, or
(ii) $W = R(a_1)R(x_1)R(a_2) \cdots R(a_l)R(x_l)$, where $a_i \in A$, $x_j \in M$.

PROOF. Let $w = R(b_1) \cdots R(b_r)$ be a rescuing operator. Then $b_r \in M$. Indeed, if $b_r \in A$, and $w_1 = R(b_1) \cdots R(b_{r-1})$ then $aw_1 \in I$ by the definition of rescuing operator. Hence $aw = aw_1 R(b_r) \in I$, a contradiction.

Now we have only to check that w does not contain two adjacent even multiplications or two adjacent odd multiplications.

Suppose that two adjacent multiplications in w have the same parity. Using the Jordan identity we can move them to the right end of w modulo a linear combination of operators that are shorter than w. Thus we can assume that b_{r-1} and b_r have the same parity. Since $b_r \in M$ it follows that $b_{r-1}, b_r \in M$. Let $w_2 = R(b_1) \cdots R(b_{r-2})$. We have $aw_2 \in I$ and therefore $aw \in I$ by Lemma 1.27, a contradiction. Lemma is proved. □

Properties of rescuing operators

Let $0 \neq a \in I$ and suppose that a rescuing operator for the element a has length r. Let $w = R(a_1)R(x_1)R(a_2) \cdots R(a_l)R(x_l)$, where $a_i \in A$, $x_j \in M$, $2l = r$. Then

(1) $aw(a_1, a_2, \dots; x_1, \dots)$ is skewsymmetric (modulo I) with respect to the even elements $a_1, a_2 \dots$.
Indeed, we have to prove that $aw(\dots, a_i, a_{i+1}, \dots) + aw(\dots, a_{i+1}, a_i \dots) \in I$. This follows from the Jordan identity applied to

$$R(a_i)R(x_j)R(a_{i+1}) + R(a_{i+1})R(x_j)R(a_i).$$

(2) aw is symmetric modulo I with respect to the odd elements $x_1, x_2, \dots$.
This follows from the Jordan identity applied to

$$R(x_i)R(a_j)R(x_{i+1}) - R(x_{i+1})R(a_j)R(x_i)$$

(3) If $a_i = a_i' a_i''$ then $aw(\dots, a_i' a_i'', \dots, a_j, \dots; \dots, x_k, \dots) =$
$aw(\dots, a_i', \dots, a_j a_i'', \dots; \dots, x_k, \dots) +$
$aw(\dots, a_i'', \dots, a_j a_i', \dots; \dots, x_k, \dots) \bmod I =$
$aw(\dots, a_i', \dots, a_j, \dots; \dots, x_k a_i'', \dots) +$
$aw(\dots, a_i'', \dots, a_j, \dots; \dots, x_k a_i', \dots) \bmod I$

Indeed, by (2) we can assume that $R(x_k)$ stands to the right of $R(a_i)$. By the Jordan identity $R(a_i' a_i'')R(x_k) = -R(x_k a_i')R(a_i'') - R(x_k a_i'')R(a_i') + R(a_i')R(x_k a_i'') + R(a_i'')R(x_k a_i') + R(x_k)R(a_i' a_i'')$.
In view of Lemma 6.3 this proves the second equality.

To prove the first equality we have to show that if $a_i = b^2$, $a_j = b$, $b \in A$, then $aw(a_1, a_2, \dots; x_1, \dots) \in I$.
Now $aw(\dots, b^2, \dots, b; x_1, \dots) = 2aw(\dots, b, \dots, b; x_1 b, \dots) \bmod I$ by (1).

(4) If at least one elements a_i lies in I then $aw \in I$.
Indeed, by (1), we can assume that $i = l$. If $a_l \in I$ then $aw \in [MI, M] \subseteq I$, by Lemma 1.27.

(5) If at least one element a_i satisfies that $\bar{a}_i$ lies in $\tilde{Z}$, then $aw \in I$.
As above we can assume that $i = l$ and then refer to Lemma 1.34.

(6) $l \leq dim(\mathcal{G})$.

Let $\{e_\alpha\}$ be a basis of $\mathcal{G}$ consisting of homogeneous elements. In view of (4) we can assume that for any i, $\bar{a}_i = e_\alpha \otimes t^r$. We will show that no α appears twice.

Indeed, suppose that $\bar{a}_i = e_\alpha \otimes t^r$, $\bar{a}_j = e_\alpha \otimes t^{r+ns}$. Let q be a preimage of t^nunder the homomorphism $A \to \mathcal{L}(G)$. Then $a_i q^s = a_j$ modulo I. Combining (3) and (5) we see that $aw(\dots, a_i, \dots, a_i q^s, \dots) \in I$. This proves the claim.

Now our aim is to prove the following proposition:

PROPOSITION 6.4. Let $\tilde{W}$ be the set of operators $R(aq) - R(a)R(q)$, $R(aq) - R(q)R(a)$, $R(u)$, where a, q u are arbitrary elements from A, $\tilde{Z}$, I respectively. Let $R = R^J < A >$. Then the ideal $id_R(\tilde{W})$ acts nilpotently on A.

For arbitrary elements $a \in A$, $q \in \tilde{Z}$, denote $w(a,q) = R(a)R(q) - R(aq)$ and $w(q,a) = R(q)R(a) - R(qa)$.

Remark that $w(a,q) - w(q,a) = D(a,q)$.

LEMMA 6.5. *1)* $w(a^2,q) = 2w(a,q)R(a)$,

2) $w(q,a^2) = 2R(a)w(q,a)$.

PROOF. 1) By the Jordan identity

$$2R(aq)R(a) + R(a^2)R(q) - R(a^2q) - 2R(a)R(q)R(a) = 0,$$

which implies the assertion (1).

The assertion (2) follows from the Jordan identity $2R(a)R(qa) + R(q)R(a^2) - R(a^2q) - 2R(a)R(q)R(a) = 0.$

Lemma is proved. □

As above by $R(A)$ we denote the vector space of right multiplications $R(a)$, $a \in A$.

Let W be the set of operators of the following types:

(1) $R(u)$ with $u \in I$,

(2) $D(a,q)$ with $a \in A$, $q \in \tilde{Z}$,

(3) $w(a,q)$, $a \in A$, $q \in \tilde{Z}$.

LEMMA 6.6. $WR \subseteq RWR(A)$.

PROOF. An arbitrary operator in R is a linear combination of operators $D_1 \cdots D_r R(b)R(b)$, where $D_i's$ are inner derivations and $b \in A$.

Since the F-linear span of W is invariant under derivations, it is sufficient to prove that $WR(b)R(b) \subseteq RWR(A)$.

Let $w \in W$. If $w = R(u)$, $u \in I$, then

$$R(u)R(b)R(b) =$$

$$-R(b)R(b)R(u) - R((ub)b) + R(u)R(b^2) + 2R(b)R(ub) \in RWR(A).$$

If $w = D(a,q)$, then:

$$D(a,q)R(b)R(b) = R(b)D(a,q)R(b) - R(bD(a,q))R(b) \in RWR(A)$$

since $bD(a,q) \in I$.

Finally, let $w = w(a,q)$. Applying linearization to Lemma 6.5 (1) we get $w(a,q)R(b) + w(b,q)R(a) = w(ab,q)$, which implies

$w(a,q)R(b)R(b) = -w(b,q)R(a)R(b) + w(ab,q)R(b)$.

Hence, $w(a,q)R(b)R(b) = -w(b,q)R(a)R(b) \mod WR(A)$.

Now, $w(b,q)R(a)R(b) = w(b,q)R(b)R(a) - w(b,q)D(b,a) = \frac{1}{2}w(b^2,q)R(a) - w(b,q)D(b,a) \in WR(A)$. Lemma is proved. □

LEMMA 6.7. $WWR(A) \subseteq RW$.

PROOF. We have $R(I)R(A) \subseteq R(A)R(I) + D(A,I) \subseteq RW$ and $D(A,\tilde{Z})R(A) \subseteq R(A)D(A,\tilde{Z}) + R(I) \subseteq RW$. Hence, if $w', w'' \in W$ and w'' is of one of the types (1),(2) then $w'w''R(A) \subseteq RW$.

Let $a, b \in A$, $q \in \tilde{Z}$. Then

$[w(b,q), R(a)] = -R(bq)R(a) + R(b)R(q)R(a) + R(a)R(bq) - R(a)R(b)R(q) = (R(b)R(q)R(a) - R(b)R(a)R(q)) + (R(b)R(a)R(q)) - R(a)R(b)R(q) + (R(a)R(bq) - R(bq)R(a)) = R(b)D(q,a) + D(b,a)R(q) + D(a,bq) = R(q)D(b,a) - R(qD(b,a)) + R(b)D(q,a) - D(bq,a)$.

If $a \in \tilde{Z}$, then $D(b,a)$, $D(q,a)$, $D(bq,a) \in W$ and $R(qD(b,a)) \in W$ because $qD(b,a) \in I$.

This implies that $[w(b,q), R(\tilde{Z})] \subseteq RW$.

Let $w'' = w(b,q)$, $a \in A$. As we have seen above, $w''R(a) = R(a)w'' + [w'', R(a)] = R(q)D(b,a) - D(bq,a) \mod RW$. Since $[W, D(A,A)] \subseteq W$ it follows that $Ww''R(a) = WR(q)D(b,a) \mod RW$. We have $R(I)R(q) \subseteq D(I,q) + R(q)R(I) \subseteq RW$. Hence $R(I)w''R(a) \in RW$.

Similarly, $D(A,\tilde{Z})R(q) \subseteq R(q)D(A,\tilde{Z}) + R(I) \subseteq RW$.

It remains only to consider an operator $w(b_1,q_1)w(b_2,q_2)R(a)$; $a, b_1, b_2 \in A$; $q_1, q_2 \in \tilde{Z}$.

We have $w(b_1,q_1)w(b_2,q_2)R(a) = w(b_1,q_1)R(a)w(b_2,q_2) + w(b_1,q_1)(R(q_2)D(b_2,a) - R(q_2D(b_2,a)) + R(b_2)D(q_2,a) - D(b_2q_2,a))$.

Let us consider each term on the right hand side separately.

Clearly, the operators $w(b_1,q_1)R(a)w(b_2,q_2)$, $w(b_1,q_1)R(q_2D(b_2,a))$, $w(b_1,q_1)R(b_2)D(q_2,a)$ lie in RW.

Now $w(b_1,q_1)R(q_2)D(b_2,a)) = R(q_2)w(b_1,q_1)D(b_2,a) + [w(b_1,q_1), R(q_2)]D(b_2,a) = R(q_2)D(b_2,a)w(b_1,q_1) + R(q_2)[w(b_1,q_1), D(b_2,a)] + D(b_2,a)[w(b_1,q_1), R(q_2)] + [[w(b_1,q_1), R(q_2)], D(b_2,a)] \in RW$, because the linear span of W is invariant with respect to all derivations and, as we have seen above, $[w(b_1,q_1), R(q_2)] \in RW$.

Similarly

$w(b_1,q_1)D(b_2q_2,a) = D(b_2q_2,a)w(b_1,q_1) + [w(b_1,q_1), D(b_2q_2,a)] \in RW$.

Lemma is proved. □

Fix k operators $w_1, \dots, w_k \in W$ and define the function

$\gamma_k : \overbrace{A \times \cdots \times A}^{k} \longrightarrow R/RW$ (quotient vector space) via

$\gamma_k(a_1, \dots a_k) = w_1R(a_1)w_2R(a_2)\cdots w_kR(a_k) + RW$

LEMMA 6.8. *γ_k is a tetrad-like function, that is, $\gamma_k(a_{\sigma(1)},\dots,a_{\sigma(k)}) = (-1)^{|\sigma|}\gamma_k(a_1,\dots,a_k)$ and $\gamma_k(a^2,a,a_3,\dots,a_k)=0$.*

PROOF. If at least one of the operators w_i is of type $D(a,q)$ then γ_k is identically zero. Indeed, let $w_i = D(a,q)$. Then $w_iR(a_i) = R(a_i)w_i - R(a_iD(a,q)) \in RW$.

Applying Lemma 6.7 several times we get:
$w_iR(a_i)w_{i+1}R(a_{i+1})\cdots w_kR(a_k) \in RW$.
Similarly, if $w_i = R(u)$, $u\in I$, then $w_iR(a_i) = R(u)R(a_i) = R(a_i)R(u) + D(u,a_i) \in RW$, since $u \in I \subseteq \tilde{Z}$. This implies that $\gamma_k = 0$.

Suppose that $a_i = a_{i+1} = a$. Let $w_{i+1} = w(b,q)$. Then, by Lemma 6.5
$R(a)w(b,q)R(a) = R(a)(-w(a,q)R(b)+w(ab,q) = -\frac{1}{2}w(a^2,q)R(b)+R(a)w(ab,q))$.

Again by Lemma 6.7 we have $w_iw(a^2,q)R(b)w_{i+2}\cdots w_kR(a_k) \in RW$ and $w(abq)w_{i+2}\cdots w_kR(a_k)\in RW$.

Now we have to prove that if $a_i = a^2$, $a_{i+1} = a$, then $\gamma_k(a_1,\dots,a_s) = 0$.
Let $w_i = w(b,q)$. Since $w(b,q) = w(q,b) + D(b,q)$ it follows that

$$w_1R(a_1)w_2R(a_2)\cdots w_kR(a_k) = w_1R(a_1)\cdots R(a^2)w(q,b)R(a)\cdots w_kR(a_k)$$

mod RW. Furthermore, $R(a^2)w(q,b) = w(q,a^2b) - R(b)w(q,a^2)$.
Clearly, $w_iw(q,a^2b)R(a_{i+1})\cdots w_kR(a_k) \in RW$.
Now, $w(q,a^2)R(a)\cdots w_kR(a_k) = 2R(a)w(q,a)R(a)\cdots w_kR(a_k)$.

Again substituting $w(q,a) = w(a,q) + D(q,a)$ and using the fact that $D(q,a)R(a) = R(a)D(q,a) - R(aD(q,a)) \in RW$ we get
$w(q,a)R(a)\cdots w_kR(a_k) = w(a,q)R(a)\cdots w_kR(a_k)$ mod RW and, finally,
$w(a,q)R(a)\cdots w_kR(a_k) = \frac{1}{2}w(a^2,q)w_{i+2}\cdots w_kR(a_k) \in RW$. Lemma is proved. □

LEMMA 6.9. *If $s \geq dim_F\mathcal{G}+1$, then $\gamma_s = 0$.*

PROOF. We can assume without loss of generality that the elements $a_i's$ are homogeneous. Let q be a preimage of t^n, $q\in A_n$. The element q is invertible. Let $a_i \in A_{s_i}$, $s_i = k_in+r_i$, $0\le r_i\le n-1$. Then $a_iq^{-k_i}\in A_{r_i}$ and $a_i-(a_iq^{-k_i})q^{k_i}\in I$.
From $s\geq dim_F\mathcal{G}+1$ it follows that the elements $a_iq^{-k_i}$, $1\le i\le s$ are linearly dependent. Hence there exists i such that $a_iq^{-k_i} = \sum_{j\neq i}\alpha_ja_jq^{-k_j}$,
$\alpha_j\in F$ and therefore $a_i \in \sum_{j\neq i}a_jF[q^{-1},q]$. In view of Lemma 6.8 this implies the result. Lemma is proved. □

COROLLARY 6.10. $(WR(A))^{dim_F\mathcal{G}+1}\subseteq RW$.

COROLLARY 6.11. $(WR)^{dim_F\mathcal{G}+1}\subseteq RW$.

PROOF. By Lemma 6.6 $WR \subseteq RWR(A)$. This implies $(WR)^{dim_F\mathcal{G}+1}\subseteq R(WR(A))^{dim_F\mathcal{G}+1}$. By Corollary 1 the right hand side of the inclusion lies in RW. Corollary is proved. □

Now we can prove Proposition 6.4. Since W and $\tilde{W}$ span the same subspace of R it is sufficient to prove that the ideal $id_R(W)$ acts nilpotently on A.

Denote $I_s = A(WR)^s$ and let us suppose that $I_s \neq (0)$ for all s.

Let $m(u)$ denote the length of the rescuing operator of an element $0 \neq u \in I$ This means that there exist elements $a_1, \dots a_{m(u)} \in A \cup M$ such that $R(a_1) \cdots R(a_{m(u)})$ is a rescuing operator for u.

Let $m_s = min\{m(u) | 0 \neq u \in I_s\}$. Consider the increasing sequence: $m_1 \leq m_2 \leq \cdots$

Since this sequence is bounded (each m_i is not greater than $2dim\mathcal{G} + 1$) there exists a number s_0 such that $m = m_{s_0} = m_{s_0+1} = \cdots$

Let $0 \neq u \in I_{s_0+dim\mathcal{G}+1}$ be an element such that $m(u) = m$. By Corollary 6.11 $u \in A(WR)^{s_0}(WR)^{dim\mathcal{G}+1} \subseteq I_{s_0}W$. Let $u = u'w$, where $u' \in I_{s_0}$, $w \in W$.

By minimality of m, the rescuing operator of u has the form

$$R(x_1)R(a_1) \cdots R(x_i) \cdots R(x_{m'}),$$

where $m = 2m' - 1$. We also know that no element of I_{s_0} can be rescued by an operator of length $\leq m - 1$.

Let $w = R(a)$, $a \in I$. By the Jordan identity
$u'R(a)R(x_1)R(a_1) = u'(-R(a_1)R(x_1)R(a)) - R(x_1(aa_1)) + R(x_1)R(aa_1) + R(a)R(x_1a_1) + R(a_1)R(x_1a))$.

Analizing each summand on the right hand side separately and using property (4) of rescuing operators we see that

$$u'R(a)R(x_1)R(a_1)R(x_2) \cdots R(x_{m'}) \in I,$$

a contradiction.

Now let $w = w(b, q) = R(b)R(q) - R(bq)$, $b \in A$, $q \in \tilde{Z}$.

Again applying the Jordan identity to the underlined part we get
$u'\underline{R(bq)R(x_1)R(a_1)} \cdots R(x_{m'}) = u'(R(x_1)R(a_1(bq)) - R(a_1)R(x_1)R(bq))R(x_2) \cdots R(x_{m'})$ mod I. (*)

Now consider the element $u'R(b)R(q)R(x_1)R(a_1)R(x_2) \cdots R(x_{m'})$. Since the element $u'b$ lies in I_{s_0} it can not be rescued by an operator of length $< m$. Hence the Jordan identity applied to the underlined part implies
$u'R(b)\underline{R(q)R(x_1)R(a_1)}R(x_2) \cdots R(x_{m'}) =$
$u'R(b)(R(x_1a_1)R(q) + R(x_1q)R(a_1) - R(a_1)R(x_1)R(q)) \cdots R(x_{m'})$ mod I.

By the property (5) of rescuing operators the operator $R(x_1)R(q)R(x_2) \cdots R(x_{m'})$ cannot rescue $u'R(b)R(a_1)$.

Hence, $u'R(b)R(a_1)R(x_1)R(q) \cdots R(x_{m'}) \in I$.

Similarly, $u'R(b)R(x_1a_1)R(q) \cdots R(x_{m'}) \in I$ since $q \in \tilde{Z}$ and $R(x_1a)R(q) \cdots$ can not rescue $u'b$.

So $u'R(b)R(q)R(x_1)R(a_1) \cdots R(x_{m'}) =$
$u'\underline{R(b)R(x_1q)R(a_1)}R(x_2) \cdots R(x_{m'}) =$
$u'(-R(a_1)R(x_1q)R(b) + R(x_1q)R(a_1b))R(x_2) \cdots R(x_{m'})$.

Arguing as above we get

$u'R(a_1)R(x_1q)R(b)R(x_2)\cdots R(x_{m'}) =$
$u'R(a_1)R(x_1)R(bq)R(x_2)\cdots R(x_{m'})$ mod I, and
$u'R(x_1q)R(a_1b)R(x_2)\cdots R(x_{m'}) = u'R(x_1)R()a_1b)q)R(x_2)\cdots R(x_{m'})$ mod I.
Since $a_1(bq) - (a_1b)q \in I$ together with (*) it implies
$u'w(b,q)R(x_1)\cdots R(x_{m'}) \in I$, a contradiction.

Finally let $w = D(b,q) = D$; $b \in A$, $q \in \tilde{Z}$.
Since $a_iD \in I$ from the property (4) of rescuing operators it follows that
$u'DR(x_1)R(a_1)\cdots R(x_{m'}) = u'R(x_1)R(a_1)\cdots R(x_{m'})D -$
$\sum_{i=1}^{m'} u'R(x_1)R(a_1)\cdots R(x_iD)\cdots R(x_{m'})$ mod I.

We have $u'R(x_1)R(a_1)\cdots R(x_{m'})D \in AD \subseteq I$. In view of property (2) of rescuing operators, to prove that all other summands also lie in I it is sufficient to prove that $u'R(x_1)R(a_1)\cdots R(x_{m'}D) \in I$.

By minimality of m the element $u'R(x_1)R(a_1)\cdots R(x_{m'-2})$ lies in I and therefore
$u'R(x_1)R(a_1)\cdots R(x_{m'-2})R(a_{m'-2})R(x_{m'-1})R(a_{m'-1})R(x_{m'}D) \in$
$[(MI)A, MD(\tilde{Z},A)] \subseteq I$ by Lemma 6.1. Proposition 6.4 is proved.

DEFINITION 6.12. We say that an ideal K of Jordan algebra A is strongly nilpotent if there exists a number s such that any product of elements of A involving s elements from K is equal to 0.

The minimal number s with this property is called the degree of strong nilpotency.

COROLLARY 6.13. The ideal I is strongly nilpotent.

PROOF. By Proposition 6.4 the ideal $id_R(\tilde{W})$ is nilpotent. Suppose that $Aid_R(\tilde{W})^s = (0)$. Then the ideal I is strongly nilpotent of degree $\leq 2s-1$. Indeed, let u be a product of elements $a_1, \ldots, a_k \in A$ and at least $2s-1$ of these elements lie in I.Then $R(u)$ is a sum of products of operators $R(a_i)$, $R(a_ia_j)$ (see [ZSSS]) and each such product contains at least s multiplications by elements from I.
Hence $R(u) \in id_R(\tilde{W})^s = (0)$. □

Now our aim will be to prove the following proposition:

PROPOSITION 6.14. Let A be a $\mathbb{Z}$-graded Jordan algebra, I a nilpotent graded ideal of A such that $A/I \simeq \mathcal{L}(G)$, a loop algebra corresponding to a simple $\mathbb{Z}/n\mathbb{Z}$-graded finite dimensional Jordan algebra of a symmetric bilinear form. Let $R = R < A >$ be the multiplication algebra of A and let $\tilde{Z}$ denote the preimage of the center of $\mathcal{L}(G)$ under the natural homomorphism $A \to \mathcal{L}(G)$. Suppose further that the ideal of R generated by the set $W = R(I) \cup D(\tilde{Z}, a) \cup \{w(q,a) = R(q)R(a) - R(qa),\ q \in \tilde{Z},\ a \in A\}$ is nilpotent. Then there exists a graded subalgebra B of A such that $A = I + B$ and $B \simeq \mathcal{L}(G)$.

Let us show that without loss of generality we can assume that $I^2 = (0)$. Indeed, we have seen above that nilpotency of the ideal $id_R(W)$ implies that the ideal I of A is strongly nilpotent. Let r be the index of strong nilpotency of I. Then the ideal $id_A(I^2)$ is strongly nilpotent of index $< r$.

If the proposition is valid for $I/id_A(I^2)$ then there exists a graded subalgebra S of A such that $I \subseteq S$, $A = I + S$ and $S/id_A(I^2) \simeq \mathcal{L}(G)$. Applying the induction assumption on the index of strong nilpotency to the ideal of S, $id_A(I^2)$, we will find a graded subalgebra $B \subseteq S$ such that $id_A(I^2) + B = S$, $S \simeq \mathcal{L}(G)$.

From now on in this section we will assume that $I^2 = (0)$.

The vector space I is an $A/I \simeq \mathcal{L}(G)$-bimodule. If N is an arbitrary graded $\mathcal{L}(G)$-subbimodule of I then N is generated as $\mathcal{L}(G)$-bimodule by $N_0 + \cdots + N_{2n-1}$. Indeed, let $F[t^{-n}, t^n]$ be the center of $\mathcal{L}(G)$ and let $q \in A_n$ be a preimage of t^n under $A \to \mathcal{L}(G)$. Then q is invertible and $q^{-1} \in A_{-n}$. This implies that as $F[U(q), U(q^{-1})]$-module N is generated by $N_0 + \cdots + N_{2n-1}$.

Hence, I satisfies the ascending chain condition and the descending chain condition for subbimodules.

Let us assume that, contrary to Proposition 6.14, the homomorphism $A \to \mathcal{L}(G)$ does not split. Let $N \leq I$ be a maximal graded $\mathcal{L}(G)$- subbimodule of I such that the homomorphism $A/N \to \mathcal{L}(G)$ does not split.

Replacing A by A/N we can assume that for any nonzero graded ideal I' of A contained in I there exists a graded subalgebra A' of A such that $I' \leq A'$, $A'/I' \simeq \mathcal{L}(G)$ and $A/I' = I/I' + A'/I'$.

Let I' be an irreducible $\mathcal{L}(G)$-subbimodule of I. Taking A' and I' instead of A and I we will assume that the $\mathcal{L}(G)$-bimodule I is irreducible.

Since $Iid_R(W)$ is a subbimodule of I and the ideal $id_R(W)$ is nilpotent, it follows that $IW = (0)$. In other words, for arbitrary elements $u \in I$, $a \in A$, $z \in F[t^{-n}, t^n]$ we have

$$u(az) = (ua)z = (uz)a.$$

The subspace $U = \sum_{i=0}^{n-1} I_i$ is a $\mathbb{Z}/n\mathbb{Z}$- graded $\mathcal{G}$-bimodule via the action:

$$u_i \star a_j = \begin{cases} u_i a_j & \text{if i+j} \leq \text{n-1} \\ (u_i a_j)t^{-n} & \text{if i+j} \geq \text{n.} \end{cases}$$

Remark that the $\mathcal{G}$-bimodule U is irreducible and that I is isomorphic to the loop bimodule, $\mathcal{L}(U)$ over $\mathcal{L}(G)$.

Let $\mathcal{G} = F1 + V$, $<,>: V \times V \to F$ is the nondegenerate symetric bilinear form. From $V = \mathcal{G}D(\mathcal{G}, \mathcal{G})$ it follows that V is a graded subspace of $\mathcal{G}$, $V = \sum_{i=0}^{n-1} V_i$.

Let $k = m - 1$ if $n = 2m$ and $k = m$ if $n = 2m + 1$. Let $\mathcal{G}' = F1 + \sum_{i=-k}^{k} V_i$ a subalgebra of $\mathcal{G}$. The subspace $F1 + \sum_{i=-k}^{k} V_i \otimes t^i$ of $\mathcal{L}(G)$ is a subalgebra which is isomorphic to $\mathcal{G}'$. Indeed, for $-k \leq i, j \leq k$ we have $V_i V_j = (0)$ unless $i + j = 0$.

LEMMA 6.15. *The algebra A contains a graded subalgebra B', $B' \subseteq \sum_{i=-k}^{k} A_i$ which is isomorphic to $\mathcal{G}'$.*

PROOF. Consider an arbitrary basis of $F1 + \sum_{i=-k}^{k} V_i \otimes t^i$ consisting of homogeneous elements. For every element of this basis choose a homogeneous preimage under the homomorphism $A \to \mathcal{L}(G)$. Let A' be a subalgebra of A generated by all these preimages. Clearly $\mathcal{G}' \simeq A'/A' \cap I$. Hence $A' \cap I$ is an ideal of finite codimension in a finitely generated algebra A'. Hence (see [ZS]) $A' \cap I$ is a finitely generated nilpotent algebra, hence $dim_F(A' \cap I) < \infty$, hence $dim_F A' < \infty$.

By the splitting theorem for graded finite-dimensional Jordan algebras (see [Ta], [J3]) there is a graded subalgebra B' in A' such that $B' \simeq \mathcal{G}'$ and $A' = B' + (A' \cap I)$. Lemma is proved. $\square$

Let us suppose that the number n is even. Let $\varphi : A \to \mathcal{L}(G)$ denote the natural homomorphism. Consider the finite dimensional Lie subalgebras $D(A_{-n/2}, A_{n/2})$ of $Der(A)$ and $D(V_{-n/2}, V_{n/2})$ of $Der(\mathcal{G})$.

Clearly $D(V_{-n/2}, V_{n/2})$ is isomorphic to the algebra of skewsymmetric transformations of $V_{n/2}$ with respect to the bilinear form $<,>$.

The homomorphism φ induces a homomorphism

$$\psi : D(A_{-n/2}, A_{n/2}) \to D(V_{-n/2}, V_{n/2})$$

of Lie algebras : if $a \in A_{-n/2}$, $b \in A_{n/2}$ and $\varphi(a) = v_{n/2} \otimes t^{-n/2}$, $\varphi(b) = v'_{n/2} \otimes t^{n/2}$ then $\psi(D(a, b)) = D(v_{n/2}, v'_{n/2})$.

LEMMA 6.16. *$Ker\, \psi = D(I_{-n/2}, A_{n/2}) + D(A_{-n/2}, I_{n/2}) + Span(D(a, b) \mid a \in A_{-n/2},\ b \in A_{n/2},\ \varphi(b) = \varphi(a)t^n)$.*

PROOF. Let $e_1, \ldots, e_r$ be an arbitrary basis of $V_{n/2}$. Then $\{D(e_i, e_j) | 1 \leq i < j \leq r\}$ is a basis of $D(V_{n/2}, V_{n/2})$.

Let $a_i \in A_{n/2}$, $b_i \in A_{-n/2}$ be preimages of the elements $e_i \otimes t^{n/2}$, $e_i \otimes t^{-n/2}$ respectively. Then $A_{n/2} = I_{n/2} + \sum_{i=1}^{r} F a_i$, $A_{-n/2} = I_{-n/2} + \sum_{j=1}^{r} F b_j$.

We have

$$D(A_{-n/2}, A_{n/2}) = D(I_{-n/2}, A_{n/2}) + D(A_{-n/2}, I_{n/2}) + \sum_{1 \leq i,j \leq r} FD(b_i, a_j)$$

and

$$\sum_{1 \leq i,j \leq r} FD(b_i, a_j) = \sum_{1 \leq i < j \leq r} FD(b_i, a_j) +$$

$$\sum_{i=1}^{r} FD(b_i, a_i) + \sum_{1 \leq i \neq j \leq r} FD(b_i + b_j, a_i + a_j).$$

Clearly $\psi(D(b_i, a_i)) = \psi(D(b_i + b_j, a_i + a_j)) = 0$, $b_i = v_i \otimes t^{n/2}$, $a_i = v_i \otimes t^{-n/2}$. By the other side, $\psi(D(b_i, a_j)) = D(e_i, e_j)$.

On the other hand the derivations $\psi(D(b_i, a_j) = D(e_i, e_j)$, $1 \leq i < j \leq r$, are linearly independent. Lemma is proved. $\square$

COROLLARY 6.17. *$AKer\psi \subseteq I$.* From $I^2 = (0)$, $IW = (0)$ it also follows that $IKer\psi = (0)$.

<u>CASE I.</u> $V_{n/2} = (0)$ or n is odd.

Since B' is a simple finite dimensional subalgebra of A it follows that A is a completely reducible B'-bimodule (see [J3]). Hence there is a graded B'-subbimodule $\mathcal{V}$ of A, $B' \subseteq \mathcal{V}$ such that $A = I + \mathcal{V}$ is a direct sum of B'-bimodules. Then the restriction $\varphi|_{\mathcal{V}} : \mathcal{V} \to \mathcal{L}(G)$ is a bijection. Let q be the preimage of t^n, $q = (\varphi|_{\mathcal{V}})^{-1}(t^n) \in \mathcal{V}_n$.

We claim that $AD(q, B') = (0)$.

Indeed, $\varphi|_{B'} : B' \to F1 + \sum_{i=-k}^{k} V_i \otimes t^i$ is an algebra isomorphism.

Choose an arbitrary element $v_{-i} \in V_{-i}$, $-k \leq i \leq k$, and let $b = \varphi|_{B'}^{-1}(v_i \otimes t^i)$. We need to show that $AD(q, b) = (0)$.

Suppose at first that $b^2 = 0$. This case automatically includes $i \neq 0$. There exists an element $v'_{-i} \in V_{-i}$ such that $< v_i, v'_{-i} >= 1$. Let $b' = \varphi|_{B'}^{-1}(v'_{-i} \otimes t^{-i})$. Then $bb' = 1$.

Since φ is a B'-bimodule homomorphism it follows that $(qb)b = 0$, $(qb)b' = q$.

Hence $D(q, b) = D((qb)b', b) = D(qb, b'b) + D(b', (qb)b) = 0$.

Now suppose that $i = 0$ and $b^2 \neq 0$. Then without loss of generality we can assume that $b^2 = 1$. As above $q = (qb)b$.

Hence, $D(q, b) = D((qb)b, b) = \frac{1}{2}D(qb, b^2) = \frac{1}{2}D(qb, 1) = 0$, which finishes the proof of the claim.

Now the subalgebra B generated by B', q, q^{-1} is isomorphic to $\mathcal{L}(G)$.

<u>CASE II</u> $dim(V_{n/2}) = 1$.

As before we choose a graded B'-subbimodule $\mathcal{V}$ of A such that $B' \subseteq \mathcal{V}$ and $A = \mathcal{V} + I$ is a direct sum. Denote $B^* = \varphi|_{B'}^{-1}(\sum_{i=-k}^{k} V_i \otimes t^i)$, $B' = F1 + B^*$.

Since $\varphi(\mathcal{V}_{n/2}B^*) \subseteq (V_{n/2} \otimes t^{n/2})(\sum_{i=-k}^{k} V_i \otimes t^i) = (0)$, it follows that $\mathcal{V}_{n/2}B^* \subseteq I \cap \mathcal{V} = (0)$.

Let $0 \neq a_{n/2} \in \mathcal{V}_{n/2}$. Let $\varphi(a_{n/2}) = v_{n/2} \otimes t^{n/2}$, $v_{n/2} \in V_{n/2}$. Since $\dim_F V_{n/2} = 1$ it follows that $< v_{n/2}, v_{n/2} > \neq 0$. Hence the elements $a_{n/2}$ is invertible and $a_{n/2}^{-1} \in A_{-n/2}$.

We will show that the subalgebra $B =< B^*, a_{n/2}, a_{n/2}^{-1} >$ is isomorphic to $\mathcal{L}(G)$.

Let us check that $a_{n/2}^2$ lies in the center of B. Indeed,
$D(a_{n/2}^2, B^*) = 2D(a_{n/2}, B^* a_{n/2}) = (0)$, $D(a_{n/2}^2, a_{n/2}^i) = 0$,
$D(a_{n/2}^2, (B^*)^2) = 0$ because $(B^*)^2 \subseteq F1$, $D(a_{n/2}^2, B^* a_{n/2}^{\pm 1}) = (0)$, because $B^* a_{n/2}^{\pm 1} = (0)$.

Now it follows that

$$B = F[a_{n/2}^2, a_{n/2}^{-2}] + B^* F[a_{n/2}^2, a_{n/2}^{-2}] + a_{n/2} F[a_{n/2}^2, a_{n/2}^{-2}] \simeq \mathcal{L}(G).$$

<u>CASE III</u> dim $V_{n/2} = 2$.

Choose elements $v, w \in V_{n/2}$ such that $< v, v >=< w, w >= 0$, $< v, w >= 1$.

Let us consider the homomorphism $\psi : D(A_{-n/2}, A_{n/2}) \to D(V_{-n/2}, V_{n/2})$ of Lie algebras, and let d be a preimage of $D(v, w)$.

The algebra A is a sum of root spaces with respect to d (because every homogeneous component A_i is d-invariant), $A = \oplus_{\alpha \in F} A^{(\alpha)}$. If $a \in A^{(\alpha)}$,

$\varphi : A \to \mathcal{L}(G)$, then $\varphi(a)$ is an eigenvector of $D(v \otimes t^{-n/2}, w \otimes t^{n/2})$ corresponding to the root α.

Clearly, v, w are eigenvectors of $D(v, w)$ that belong to the eigenvalues -1,1 respectively.

Let us consider the root space $A^{(0)}$ which is a graded subalgebra of A. Then $\varphi((A^{(0)}) = F1 + \sum_{i \neq \frac{n}{2}+n\mathbb{Z}} V_{\bar{i}} \otimes t^i$.

By Lemma 6.15 we can choose a graded subalgebra B' of $A^{(0)}$ such that $\varphi|_{B'} : B' \to F1 + \sum_{i=-(\frac{n}{2}-1)}^{\frac{n}{2}-1} V_{\bar{i}} \otimes t^i$ is an isomorphism. Let $B^* = \varphi|_{B'}^{-1}(\sum_{i=-(\frac{n}{2}-1)}^{\frac{n}{2}-1} V_{\bar{i}} \otimes t^i)$, $B' = F1 + B^*$.

Since $B' \subseteq A^{(0)}$ every root space $A^{(\alpha)}$ is a completely reducible bimodule over B'. Hence, there exists a B'-subbimodule $\bar{A}^{(\alpha)}$ of $A^{(\alpha)}$ such that $A^{(\alpha)} = \bar{A}^{(\alpha)} + (I \cap A^{(\alpha)})$ is a direct sum.

Choose elements $a \in \bar{A}^{(-1)}_{n/2}$, $b \in \bar{A}^{(1)}_{n/2}$ such that $\varphi(a) = v \otimes t^{n/2}$, $\varphi(b) = w \otimes t^{n/2}$. The element a^2 lies in $A^{(-2)}$. On the other hand $v^2 = 0$ implies that $a^2 \in I$.

Recall that I is a Jordan bimodule over A/I and thus a module over the Lie algebra $D(\mathcal{L}(G), \mathcal{L}(G))$.

LEMMA 6.18. *I has no nonzero eigenvectors of $D(v \otimes t^{-n/2}, w \otimes t^{n/2})$ that belong to the eigenvalue -2.*

PROOF. As we have noticed above a $\mathcal{L}(G)$-bimodule I is isomorphic to a loop bimodule $\mathcal{L}(U)$ that corresponds to the $\mathbb{Z}/n\mathbb{Z}$-graded finite dimensional bimodule U over $\mathcal{G}$.

The bimodule U is a direct sum of irreducible $\mathcal{G}$-bimodules. Every irreducible $\mathcal{G}$-bimodule is contained in a bimodule of r-vectors (see [J3]). The eigenvalues of $D(v, w)$ in a bimodule of r-vectors are -1,0,1. This proves the lemma. □

COROLLARY 6.19. $a^2 = b^2 = 0$.

Denote $q = ab$. Then $\varphi(q) = t^n$. We will prove that $B =< B^*, a, b, q^{-1} >\simeq \mathcal{L}(G)$. As in case II $B^* a \subseteq B^* \bar{A}^{(-1)}_{n/2} \subseteq I \cap \bar{A}^{(-1)} = (0)$ and similarly $B^* b = (0)$.

Let us check that q lies in the center of B.

Indeed, $D(q, a) = D(ab, a) = \frac{1}{2} D(b, a^2) = 0$ and similarly $D(q, b) = (0)$; $D(q, B^*) = D(ab, B^*) = D(a, bB^*) + D(b, aB^*) = (0)$.

Since $B^* B^* \subseteq F1$ it follows that $D(q, B^* B^*) = (0)$. It remains to show that $D(q, B^* q^{-1}) = 0$ and $D(q, aq^{-1}) = D(q, bq^{-1}) = 0$.

Notice that $D(q, B^* q^{-1}) = D(qB^*, q^{-1})$ and $D(q, aq^{-1}) = D(qa, q^{-1})$, $D(q, bq^{-1}) = D(qb, q^{-1})$.

Furthermore, $R(q^{-1}) = R(q)U(q^{-1}) = R(q)U(q)^{-1}$.

That's why to show that $R(qB^*)$ commutes with $R(q^{-1})$ it is sufficient to show that $R(qB^*)$ commutes with $R(q)$ and with $U(q) = 2R(q)^2 - R(q^2)$.

We have $D(qB^*, q) = D(qB^*, ab) \subseteq D((qB^*)a, b) + D((qB^*)b, a)$. From $D(q, a) = 0$ it follows that $(qB^*)a = q(B^* a) = (0)$. Similarly, $(qB^*)b = (0)$.

Now,

$$D(qB^*, q^2) = D((qB^*)q, q) \subseteq D(B^*R(q)^2R(a), b) + D(B^*R(q)^2R(b), a).$$

As above we conclude that $B^*R(q)^2R(a) = B^*R(a)R(q)^2 = (0)$ and $B^*R(q)^2R(b) = (0)$.

Let us show that $R(qa)$ commutes with $R(q)$ and with $R(q^2)$. We have

$$D(qa, q) = D(qa, ab) = D((qa)a, b) + D((qa)b, a) =$$

$$D(qa^2, b) + D(q^2, a) = D(q^2, a) = 2D(q, qa) = -2D(qa, q).$$

Since the characteristic of F is 0 we get $D(aq, q) = 0$.

Now, $D(qa, q^2) = D(q, aq^2) + D(a, q^3)$. By the Jordan identity $R(aq^2) = -2R(q)R(a)R(q) + 2R(aq)R(q) + R(q^2)R(a)$. The operator $R(q)$ commutes with every factor on the right hand side. Similarly, $R(q^3) = -2R(q)^3 + 3R(q^2)R(q)$. The operator $R(a)$ commutes with $R(q)$ and with $R(q^2)$.

Similarly $D(qb, q) = D(qb, q^2) = 0$.

We showed that q lie in the center of B and therefore

$$B = F[q^{-1}, q] + B^*F[q^{-1}, q] + aF[q^{-1}, q] + bF[q^{-1}, q] \simeq \mathcal{L}(G).$$

<u>CASE IV</u> $dim_F(V_{n/2}) \geq 3$.

Now the Lie algebra $D(V_{n/2}, V_{n/2})$ is simple. Hence there is a subalgebra $L \subseteq D(A_{-n/2}, A_{n/2})$ such that $L \simeq D(V_{n/2}, V_{n/2})$ and $D(A_{-n/2}, A_{n/2}) = Ker\psi + L$.

Let $A^{(0)} = \{a \in A | aL = (0)\}$. Clearly, $\varphi(A^{(0)}) = Ker_{\mathcal{L}(G)} D(V_{n/2} \otimes t^{-n/2}, V_{n/2} \otimes t^{n/2}) = F1 + \sum_{i \notin \frac{n}{2} + n\mathbb{Z}} V_i \otimes t^i$. Choose a graded subalgebra B' of $A^{(0)}$ such that $\varphi|_{B'} : B' \to F1 + \sum_{i=-(\frac{n}{2}-1)}^{\frac{n}{2}-1} V_i \otimes t^i$ is an isomorphism.

Let e be a root element of L with respect to some Cartan subalgebra. Since every homogeneous component A_i is an L-module of dimension $\leq d$ it follows that $A_i e^d = (0)$. Hence $Ae^d = (0)$.

From representation theory of finite dimensional semisimple Lie algebras (see [J2]) it follows that the subalgebra of $End_F A$ generated by L is a finite dimensional semisimple algebra.

The subalgebra of $End_F A$ generated by $R(B')$ is also finite dimensional and semisimple (see J1]) and besides $[R(B'), L] = (0)$. Hence the subalgebra generated by $R(B')$ and L is finite dimensional and semisimple.

Hence there exists a graded subspace $A' \leq A$ such that $A = I + A'$ is a direct sum of vector spaces, $A'B' \subseteq A'$ and $A'L \subseteq A'$.

<u>Case IV.1.</u> $dim_F(V_{n/2}) = 2m$ is even.

In this case the algebra $D(V_{n/2}, V_{n/2})$ is of type D_m. Choose a basis in $V_{n/2}$ such that $< v_i, v_{m+i} >= 1$, $1 \leq i \leq m$ and all the other products are equal to 0.

The restriction $\varphi|_{A'} : A' \to \mathcal{L}(G)$ is a bijection. Denote $a_i = \varphi|_{A'}^{-1}(v_i \otimes t^{n/2}) \in A'_{n/2}$. The L-module $A'_{n/2}$ can be identified with the $D(V_{n/2} \otimes t^{-n/2}, V_{n/2} \otimes t^{n/2})$-module $V_{n/2} \otimes t^{n/2}$ via φ.

The elements $D(v_i, v_{m+i})$, $1 \le i \le m$ span a Cartan subalgebra of $D(V_{n/2}, V_{n/2})$. Let $h_i = \psi|_L^{-1} D(v_i, v_{m+i})$, $1 \le i \le m$.

The element v_i (resp. a_i), $1 \le i \le m$, belongs to the root $(0, \dots, -1, 0, \dots, 0)$ with respect to the above mentioned Cartan subalgebra.

The element v_{m+i} (resp. a_{m+i}), $1 \le i \le m$, belongs to the root $(0, \dots, 1, 0, \dots, 0)$.

The element a_i^2 belongs to the root $(0, \dots, -2, \dots, 0)$ and therefore $a_i^2 \in I$. Since no finite dimensional Jordan $\mathcal{G}$-bimodule has the root $(0, \dots, -2, \dots, 0)$ with respect to $D(V_{n/2}, V_{n/2})$ it follows that $a_i^2 = 0$.

Let $1 \le i < j \le m$. The basis $\{v_1, \dots, v_{i-1}, v_i + v_j, v_{i+1}, \dots, v_m, v_{m+1}, \dots, v_{m+j-1}, v_{m+j} - v_{m+i}, v_{m+j+1}, \dots, v_{2m}\}$ has the same property as $\{v_1, \dots, v_{2m}\}$.

We have $(\varphi|_{A'})^{-1}(v_i + v_j) = a_i + a_j$. Hence $(a_i + a_j)^2 = 0$, $a_i a_j = 0$. Similarly $a_{m+i} a_{m+j} = 0$ for arbitrary $1 \le i, j \le m$.

Now again choose arbitrary $1 \le i \ne j \le m$.

We have $v_{m+j} = v_i D(v_{m+i}, v_{m+j})$.

Hence $a_{m+j} = a_i D_0$, where $D_0 = (\psi|_L)^{-1} D(v_{m+i}, v_{m+j})$. We have $0 = a_i^2 D_0 = 2a_i a_{m+j} = 0$.

Now let $D = (\psi|_L)^{-1}(D(v_1, v_2) + D(v_{m+1}, v_{m+2}))$.

We have $a_1 D = a_{m+2}$, $a_{m+1} D = a_2$, $a_{m+2} D = -a_1$, $a_2 D = -a_{m+1}$. Hence $a_1 D^2 = -a_1$, $a_{m+1} D^2 = -a_{m+1}$.

Notice that $(a_1 a_{m+1}) D = a_{m+2} a_{m+1} + a_1 a_2 = 0$. Therefore, $0 = (a_1 a_{m+1}) D^2 = a_1 D^2 a_{m+1} + a_1 a_{m+1} D^2 + 2(a_1 D)(a_{m+1} D) = -2a_1 a_{m+1} + 2a_2 a_{m+2}$.

In the same way we can prove that $a_1 a_{m+1} = a_2 a_{m+2} = \cdots = a_m a_{2m}$.

Denote $q = a_i a_{m+1}$. The element q is a preimage of t^n with respect to φ and thus is invertible.

We will show that $B = < B', a_1, \cdots a_{2m}, q^{-1} > \simeq \mathcal{L}(G)$. As above it is sufficient to prove that q lies in the center of B.

We have $D(B^*, q) \subseteq D(B^*, a_1 a_{m+1}) = D(B^* a_1, a_{m+1}) + D(B^* a_{m+1}, a_1) = (0)$ and $D(a_i, q) = D(a_i, a_i a_{m+i}) = \frac{1}{2} D(a_i^2, a_{m+i}) = 0$ for $1 \le i \le m$.

Similarly, $D(a_{m+i}, q) = \frac{1}{2} D(a_{m+i}^2, a_i) = 0$.

Now it remains to check that $D(B^* q^{-1}, q) = (0)$ and $D(a_i q^{-1}, q) = 0$. This can be done exactly in the same way as in case III.

<u>Case IV. 2.</u> $dim_F(V_{n/2}) = 2m + 1$.

Then $D(V_{n/2}, V_{n/2})$ is of type B_m.

Again we choose a basis $\{v_1, v_2, \dots, v_{2m+1}\}$ in $V_{n/2}$ such that $< v_i, v_{m+i} >= 1$, $1 \le i \le m$, $< v_{2m+1}, v_{2m+1} >= 1$ and all the other products are equal to zero. Let $a_i = (\varphi|_{A'})^{-1}(v_i \otimes t^{n/2})$, $1 \le i \le 2m + 1$.

In the same way as above we can prove that $a_i a_j = 0$ for $1 \leq i, j \leq 2m$ unless $|i-j| = m$ and that $a_1 a_{m+1} = \cdots = a_m a_{2m}$. For $1 \leq i \leq m$ we have $a_i \psi|_L^{-1}(D(v_{m+i}, v_{2m+1})) = a_{2m+1}$, which implies $a_i a_{2m+1} = 0$. Similarly $a_{m+i} a_{2m+1} = 0$.

Let $D = (\psi|_L)^{-1}(D(v_{m+1}, v_{2m+1}) + D(v_1, v_{2m+1}))$.
Then $a_1 D = a_{2m+1}$, $a_{m+1} D = a_{2m+1}$, $a_1 D^2 = a_{m+1} D = -a_{m+1} - a_1 = a_{m+1} D^2$. Hence, $(a_1 a_{m+1})D = 0$ and therefore $(a_1 a_{2m+1})D^2 = (a_1 D^2) a_{m+1} + 2(a_1 D)(a_{m+1} D) + a_1(a_{m+1} D^2) = -2a_1 a_{m+1} + 2a_{2m+1}^2 = 0$.

We proved that $q = a_1 a_{m+1} = \cdots = a_m a_{2m} = a_{2m+1}^2$.
Now in the same way as before it follows that

$$B = < B', a_1, \cdots, a_{2m+1}, q^{-1} > \simeq \mathcal{L}(G).$$

Proposition 6.4 is proved.

6.2. Structure of J

In what follows we assume that the even part $A = I + \mathcal{L}(G)$ is a semidirect sum of a loop algebra $\mathcal{L}(G)$ corresponding to a finite dimensional simple Jordan algebra $\mathcal{G} = F1 + V$ of a symmetric bilinear form in a $\mathbb{Z}/n\mathbb{Z}$-graded vector space $V = V_0 + \cdots + V_{n-1}$, $dim V \geq 2$, $\mathcal{L}(G) = F[t^{-n}, t^n] + \sum_{i=j \, mod \, n} V_i \otimes t^j$, I is the nilpotent radical of A, $I \neq (0)$.

Denote $V^\sharp = \sum_{i=j \, mod \, n} V_i \otimes t^j$, $Z = F[t^{-n}, t^n]$ and $M_I = (MI)A$. Clearly M_I is an A-subbimodule of M.

LEMMA 6.20. $[M_I, M] \subseteq I + V^\sharp$.

PROOF. Suppose the contrary. Then we can choose homogeneous elements $x, y \in M$, $u \in I$, $a \in A$ such that

$$[(xu)a, y] = \alpha t^{nk} + v + u',$$

where $0 \neq \alpha \in F$, $v \in V^\sharp$, $u' \in I$. Hence, $t([(xu)a, y]t^{-nk}) = \alpha \neq 0$. From the Jordan identity we get

$$[(xu)a, y]t^{-nk} = -[(xu)t^{-nk}, y]a - [xu, y(at^{-nk})] + [xu, ay]t^{-nk} + [xu, t^{-nk}y]a + [(xu)(at^{-nk}), y].$$

By Lemmas 1.27 and 1.34 all summands except for the last one lie in I whereas the last summand has zero trace by Lemma 1.28, a contradiction. Lemma is proved □

LEMMA 6.21. $V^\sharp \subseteq [M_I, M] + I$.

PROOF. Since $I \neq (0)$ it follows that $[M_I, M] \not\subseteq I$ for otherwise $id_J(I) \cap A = I$. The nonzero subspace $[M_I, M] + I/I$ lies in $V^\sharp + I/I$ and is invariant with respect to all derivations of A/I. Since $V^\sharp + I/I$ is a graded irreducible module over $\mathrm{Der}(A/I)$ we have $[M_I, M] + I = V^\sharp + I$, which implies the lemma. □

LEMMA 6.22. *$[((MV^\sharp)I)A, M] \subseteq I$, or equivalently, $[(MI)A, MV^\sharp] \subseteq I$.*

PROOF. In view of Lemma 6.21 it is sufficient to prove that

$$[((M[(MI)A, M])I)A, M] \subseteq I.$$

Choose arbitrary elements $x, y, x', y' \in M$; $u, u' \in I$; $a, b \in V^\sharp$. We need to verify that

$$uR(x)R(a)R(y)R(x')R(u')R(b)R(y') \in I.$$

By the Jordan identity

$R(y)R(x')R(u') = R(u')R(x')R(y) - R([yu', x']) + R(y)R(x'u') - R(x')R(yu') + R(u')R([y, x'])$.

We will have to consider all five summands on the right hand side separately:

(1) $uR(x)R(a)R(u')\underline{R(x')R(y)R(b)}R(y') =$
$uR(x)R(a)R(u')(R(b)R(y)R(x') - R([x'b, y]) + R([x', y])R(b) +$
$R(x'b)R(y) - R(yb)R(x'))R(y') \in [(((MI)A)I)A, M]R(M)R(M) +$
$[(((MI)A)I)R<A>, M] \subseteq I$,

(2) $uR(x)R(a)R([yu', x'])R(b)R(y') \in [(((MI)A)I)A, M] \subseteq I$, by Proposition 1.43(3).

(3) $uR(x)R(a)R(y)R(x'u')R(b)R(y') = (x'u')R(w)R(b)R(y')$, where $w = uR(x)R(a)R(y) \in V^\sharp + I$ by Lemma 6.20. Applying the Jordan identity to $R(w)R(b)R(y')$, we get $(x'u')R(w)R(b)R(y') = (x'u')(-R(y')R(b)R(w) - R((wy')b) + R(wb)R(y') + R(wy')R(b) + R(by')R(w)) \in [MI, M]R(A)R(A) + (x'u')R(wb)R(y')$.

The first summand lies in I by Lemma 1.27. As for the second summand, we have $wb \in Z + I$. Hence $(x'u')R(wb)R(y') \in I$ by Lemma 1.34.

(4) $uR(x)R(a)R(x')R(yu')R(b)R(y')$ can be shown to lie in I in the same way as in (3).

(5) $uR(x)R(a)R(u')R([y, x'])R(b)R(y') \in [(((MI)A)I)R(A)R(A), M] \subseteq I$.
Lemma is proved. □

LEMMA 6.23. *Let $x, y \in M$; $u \in I$, $v \in V^\sharp$ be homogeneous elements such that $[(xu)v, y] \notin I$. Then M is generated by x and y as an A-bimodule.*

PROOF. For an arbitrary inner derivation $D \in D(V^\sharp, V^\sharp)$ we have

$$[(xu)v, y]D = [(xuD)v, y] + [(xu)vD, y] + [((xD)u)v, y] + [(xu)v, yD]$$

By Lemma 6.22, $[((xD)u)v, y] + [(xu)v, yD] \in I$.

Since $V^\sharp + I/I$ is a graded irreducible $Der(A/I)$-module it follows that $V^\sharp \subseteq [(xI)V^\sharp, y] + I$.

Remark that by Lemma 1.27 $[(xI)V^\sharp, y] \subseteq V^\sharp D(xI, y) + I$, and that $MD(xI, y) \subseteq [M, xI]y + [M, y](xI) \subseteq xR<A> + yR<A>$.

This implies that $M(V^\sharp D(xI, y)) \subseteq xR<A> + yR<A>$.

Hence, for an arbitrary element $v \in V^\sharp$ there exists an element $n(v) \in I$ such that $M(v - n(v)) \subseteq xR < A > +yR < A >$.

Choose an arbitrary element $z \in M$ and two homogeneous elements $v', v'' \in V^\sharp$ such that $v'v'' = 1$. We have
$z = 1z = v'R(v'')R(z) \in v'R([(xI)V^\sharp, y])R(z) + v'R(I)R(z) \subseteq$
$v'R(z)R(A) + v'D([(xI)V^\sharp, y], z) + zI$.

Remark that $D([(xI)V^\sharp, y]), z) \subseteq D((xI)V^\sharp, [y, z]) +$
$D(y, [(xI)V^\sharp, z]) \subseteq D(xR < A >, A) + D(y, A)$ and therefore
$AD([(xI)V^\sharp, y], z) \subseteq xR < A > +yR < A >$.

Hence, there exist elements $a \in A$ and $u' \in I$ such that $z = zR(n(v'))R(a) + zR(u')$ mod $xR < A > +yR < A >$.

The operator $R(n(v'))R(a) + R(u')$ is nilpotent (see [Sk]). Hence, $z \in xR < A > +yR < A >$. Lemma is proved. □

LEMMA 6.24. *Let $x, y \in M$ be homogeneous elements such that $[(xI)V^\sharp, y] \not\subseteq I$. Then $[(xI)V^\sharp, x] \not\subseteq I$.*

PROOF. Choose homogeneous elements $u \in I$; $v, w \in V^\sharp$ such that $[(xu)v, y] = w \, mod \, I$ and $w \neq 0$. Then $vw = 0$ because $[(xu)v, y]v \in I$ by the Jordan identity.

Let us check that $[(xI)w, y] \not\subseteq I$. Indeed, there exists an element $w' \in V^\sharp$ such that $ww' = 1$. We have $wD(w', v) = v$ and therefore $[(xu)v, y] = [(xu)w, y]D(w', v) -$
$[(x(uD(w', v)))w, y] - [(xD(w', v)u)w, y] -$
$[(xu)w, yD(w', v)]$.

The last two summands lie in I by Lemma 6.22.

If $[(xI)w, y] \subseteq I$ then $[(xu)v, y] \in I$, a contradiction. Thus,

$$[(xI)w, y] = uR(x)R(v)\underline{R(y)R(Ix)}R(y) \not\subseteq I.$$

We have $R(y)R(Ix) \subseteq R(I)R([y, x]) + R(x)R(yI) + R(Ix)R(y) + R(Iy)R(x) + R([x, y])R(I)$.

We will consider each summand on the right hand side separately:

(1) $uR(x)R(v)R(I)R([y, x])R(y) \subseteq [(((MI)A)I)A, M] \subseteq I$;

(2) $uR(x)R(v)R(x)R(yI)R(y)$ lies in I if we suppose that $uR(x)R(v)R(x) \in I$;

(3) $uR(x)R(v)R(Ix)R(y)R(y) \subseteq [M, Ix]R(y)^2 \subseteq I$;

(4) $uR(x)R(v)R(Iy)R(x)R(y) \subseteq [M, Iy]R(x)R(y) \subseteq I$;

(5) $uR(x)R(v)R([x, y])R(I)R(y) \subseteq [MI, y] \subseteq I$.

Hence $[(xu)v, x] \notin I$. Lemma is proved. □

In chapter 4 we have already constructed involutions in $V^\sharp$. Let's recall the construction. Choose a basis in the vector space V_0 consisting of pairwise orthogonal involutions. If $1 \leq i \leq n-1$ and $i \neq \frac{n}{2}$ choose dual bases $\{v_{i,q}\}_q$ in V_i and $\{v_{n-i,q}\}_q$ in V_{n-i}, so $< v_{i,q}, v_{n-i,l} >= \delta_{ql}$. For each pair $v_{i,q}, v_{n-i,q}$ the elements $\sqrt{\frac{1}{2}}(v_{i,q} \otimes t^i + v_{n-i,q} \otimes t^{-i})$, $\sqrt{\frac{-1}{2}}(v_{i,q} \otimes t^i - v_{n-i,q} \otimes t^{-i})$ from $V^\sharp$ are involutions.

If $\dim V_{n/2}$ is even, then $V_{n/2} = V'_{n/2} \oplus V''_{n/2}$, where $V'_{n/2}$, $V''_{n/2}$ are isotropic dual subspaces.

If $\dim V_{n/2}$ is odd, then $V_{n/2} = Fw \oplus V'_{n/2} \oplus V''_{n/2}$, where $< w, w >= 1$, w is orthogonal to $V'_{n/2}$ and $V''_{n/2}$, $V'_{n/2}$, and $V''_{n/2}$ are isotropic dual subspaces. In $V'_{n/2}$, $V''_{n/2}$ we again choose dual bases and form $\dim(V'_{n/2} \oplus V''_{n/2})$ involutions in the same way as above.

If $\dim V_{n/2}$ is even, then our construction yields a system of involutions $v_1, \dots ,$ $v_d \in V^\sharp$ such that $v_i v_j = \delta_{ij}$, $1 \le i, j \le d$, $d = dimV$.

If $\dim V_{n/2}$ is odd, then we get $d-1$ involutions $v_1, \dots , v_{d-1}$ and the element $v_d = w \otimes t^{n/2}$. Still $v_i v_j = 0$ for $i \neq j$, $v_i^2 = 1, 1 \le i \le d-1$, $v_d^2 = t^n$. In both cases $V^\sharp = \sum_{i=1}^d v_i Z$.

REMARK 6.25. (see Chapter 4) $D(V^\sharp, Z) = (0)$. Indeed, for an arbitrary $1 \le i \le d$ choose $1 \le j \le d$, $j \neq i$. We have $Z = (v_j Z)^2$. Hence $D(v_i Z, (v_j Z)^2) \subseteq D((v_i Z)(v_j Z), v_j Z) = (0)$.

Now we will consider in greater detail the second case when $v_d^2 = t^n$. Denote $U = U(v_d, v_d^{-1}) = 2R(v_d)R(v_d^{-1}) - Id$. We have

$$U^2 = \frac{1}{2}Id + \frac{1}{2}U(v_d^2, v_d^{-2}) = R(t^n)R(t^{-n}).$$

Hence $U^2 - Id = R(t^n)R(t^{-n}) - R(t^n t^{-n})$.

From Proposition 6.4 it follows that the operator $U^2 - Id : A \to A$ is nilpotent and therefore $U : A \to A$ has only eigenvalues ± 1.

LEMMA 6.26. *Let $a \in J$ be a root element of U. Then*

(1) a belongs to the eigenvalue -1 if and only if there exists $s \ge 1$ such that $aR(v_d)^s = 0$,

(2) a belongs to the eigenvalue 1 if and only if the following sequence: $a_1 = a, \dots , a_{n+1} = v_d D(a_n, v_d)$ vanishes.

PROOF. (1) $U + Id = 2R(v_d)R(v_d^{-1}) = 2R(v_d)^2 U(v_d^{-1})$. Hence $a(U + Id)^k = 0$ if and only if $aR(v_d)^{2k} = 0$.

(2) We have $a(U - Id) = 2(aR(v_d^{-1})R(v_d) - a) = 2v_d^{-1} D(a, v_d)$. Define a new sequence $a'_1 = a, \dots , a'_{n+1} = v_d^{-1} D(a'_n, v_d)$. Clearly, $a(U - Id)^k = 0$ if and only if $a'_{k+1} = 0$. We will use induction to prove that $a'_n U(v_d)^{n-1} = (-1)^{n-1} a_n$. For n = 1 the assertion is clear. By definition $a'_{n+1} = v_d^{-1} D(a'_n, v_d) = a'_n(R(v_d^{-1})R(v_d) - Id)$. Applying $U(v_d)^n$ to both sides, we get $a'_{n+1} U(v_d)^n = a'_n U(v_d)^{n-1}(R(v_d^{-1})R(v_d) - Id)U(v_d) = v_d^{-1} D(a'_n U(v_d)^{n-1}, v_d)U(v_d) = (-1)^{n-1} v_d^{-1} D(a_n, v_d) U(v_d)$ and it remains to notice that for an arbitrary derivation D we have $(v_d^{-1} D)U(v_d) = -v_d D$. This proves the lemma. □

LEMMA 6.27. *Let $a, b \in J$ be root elements of U belonging to eigenvalues α, β $(= \pm 1)$ respectively. Then the product ab is a root element belonging to $\alpha\beta$.*

PROOF. Suppose that $\alpha = -1$, $\beta = 1$. Then $(ab)R(v_d)^2 = (av_d)R(b)R(v_d) + aD(b, v_d)R(v_d) = ((av_d)b)R(v_d) + (av_d)D(b, v_d) - aR(v_dD(b, v_d)) = 2((av_d)b)R(v_d) - (aR(v_d)^2)b - ab_2$, where $b_2 = v_dD(b, v_d)$. As above, let $b_{i+1} = v_dD(b_i, v_d)$, $i \geq 1$ Now if $aR(v_d)^s = 0, b_q = 0$,then $(ab)R(v_d)^{2(s+q)} = 0$.

Suppose that $\alpha = \beta = -1$. Then $v_dD(ab, v_d) = v_dD(a, bv_d) + (-1)^{|a||b|}v_dD(b, av_d) = (v_da)(bv_d) - (-1)^{|a||b|}((bv_d)v_d)a + (-1)^{|a||b|}(bv_d)(av_d) - ((av_d)v_d)b$.

If $aR(v_d)^s = bR(v_d)^q = 0$ then $(ab)_{s+q-1} = 0$.

Suppose now that $\alpha = \beta = 1$. The following identity holds in an arbitrary Jordan algebra

$$cD(cD(ab, c), c) = -(cD(cD(a, c), c))b - a(cD(cD(b, c), c))+$$
$$2cD((cD(a, c))b, c) + 2cD(a(cD(b, c)), c) + 2(cD(a, c))(cD(b, c)).$$

Since the identity has degree 6 it is sufficient to check it only in special Jordan algebras which is straightforward (see [J3], Glennie identities).

Now coming back to root elements $a, b \in J$ and assuming $a_s = b_q = 0$ we get $(ab)_{2(s+q-1)} = 0$. Lemma is proved. □

Let G be a subsemigroup of $End_F(J)$ generated by $U(v_1), \dots , U(v_{d-1})$, $U(v_d, v_d^{-1})$ (if $v_d^2 = 1$ then $U(v_d, v_d^{-1}) = U(v_d)$). Clearly, G is abelian and therefore both A and M decompose into direct sums of root spaces with respect to the action of G.

In what follows, by a root element we mean a root element with respect to G. A root element u belongs to a root $(\alpha_1, \dots , \alpha_d)$ if for any i, $1 \leq i \leq d$, there exists $r_i \geq 1$ such that $x(U(v_i, v_i^{-1}) - \alpha_i Id)^{r_i} = 0$.

LEMMA 6.28. *There exists a root element $x \in M$ belonging to the root $(-1, -1, \dots , -1)$ such that $[(xI)V^\sharp, x] \nsubseteq I$ and therefore $M = xR < A >$.*

PROOF. Since $I \neq (0)$ there exist root elements $x, y \in M$ such that $[(xI)V^\sharp, y] \nsubseteq I$. Suppose that $x(U(v_i, v_i^{-1}) - \alpha_i Id)^r = x(2R(v_i)R(v_i^{-1}) - (1 + \alpha_i)Id)^r = 0$.

If $\alpha_i \neq -1$ then $x \in MV^\sharp$ which contradicts Lemma 6.22. Hence x belongs to the root $(-1, -1, \dots , -1)$. Now it remains to refer to Lemma 6.24. Lemma is proved. □

COROLLARY 6.29. $U(v_d, v_d^{-1}) : M \to M$ has only eigenvalues ± 1.

REMARK 6.30. $M = xR < Z > +MV^\sharp + M_I$. Indeed, M is spanned by elements $y = xR(u_1) \cdots R(u_k)$, where $u_i \in Z \cup V^\sharp \cup I$, $1 \leq i \leq k$. If at least one element u_i lies in I then $y \in M_I$. Suppose that $u_i \in V^\sharp$ and i is maximal with this property. Since $D(V^\sharp, Z) = (0)$ it follows that $R(u_i)$ can be moved to the right end, so $y \in MV^\sharp$. If $u_i \in Z$ for all i then $y \in xR < Z >$.

Indeed, the root element x belongs to the eigenvalue -1 with respect to $U(v_d, v_d^{-1})$ and the only eigenvalues of $U(v_d, v_d^{-1}) : A \to A$ are ± 1. Now the assertion follows from Lemma 6.27

COROLLARY 6.31. The operator $U(v_d, v_{d-1})$ is invertible and thus G is a group.

Let W be the set of multiplication operators of the following types:

(1) $R(x)R(v_{i_1})R(x)\cdots R(x)R(v_{i_{2r}})$, where $1 \leq i_1 < \cdots < i_{2r} \leq d$,

(2) $R(v_{i_1})R(x)R(v_{i_2})R(x)\cdots R(x)R(v_{i_{2r+1}})$, where $1 \leq i_1 < \cdots < i_{2r+1} \leq d$.

Clearly, $|W| < \infty$.

Define the mapping $tr : A \to Z$, by $tr(a + v + u) = a$, where $a \in Z$, $v \in V^\sharp$, $u \in I$.

LEMMA 6.32. *For an arbitrary nonzero element $u \in I$ there exists an operator $w \in W$ such that $tr(uw) \neq 0$.*

PROOF. Recall that in the first section of this chapter by a rescuing operator for an element $u \neq 0$ we meant a product P of multiplications such that $uP \in A - I$ and P has the minimal length among all operators with these properties. In section 1 we showed that a rescuing operator P:

(i) has an odd multiplication at the right end,

(ii) does not contain two subsequent even multiplications,

(iii) does not contain two subsequent odd multiplications.

Thus, $P = R(a_1)R(x_1)\cdots R(a_k)R(x_{2r})$, $k = 2r$ or $P = R(x_1)R(a_1)R(x_2)\cdots R(a_k)R(x_{2r})$, $k = 2r-1$, $a_i \in A$, $x_j \in M$.
The expression $uP + I/I$ is skew-symmetric in a_i and symmetric in x_j.
If at least one a_i lies in $I + Z$ then $uP \in I$ (see section 1).

Since $A = I+Z+V^\sharp$ all elements a_i can be assumed to be of the type $v_l t^{nj}$, $1 \leq l \leq d$. Now, $R(v_l t^{nj})R(x_{2r}) = R(x_{2r})R(v_l t^{nj}) + R(v_l)R(x_{2r}t^{nj}) - R(x_{2r}t^{nj})R(v_l) + R(t^{nj})R(x_{2r}v_l) - R(x_{2r}v_l)R(t^{nj})$.

Hence, $u\cdots R(x_{2r-1})R(v_l t^{nj})R(x_{2r}) = u\cdots R(x_{2r-1})R(v_l)R(x_{2r}t^{nj})$ *mod* I.

Now without loss of generality we will assume that an arbitrary a_i lies in $\{v_1, \ldots, v_d\}$.

If at least one element x_j lies in M_I then we can assume that $x_{2r} \in M_I$, which implies $uP \in [(Ix_{2r-1})a_k, M_I] \subseteq I$ by Lemma 1.41.
If $x_{2r} \in MV^\sharp$ then by Lemma 6.22 $uP \in I$. Since $M = M_I + MV^\sharp + xR < Z >$, we will assume that $x_{2r} \in xR < Z >$.

Let $x_{2r} = x'f$, $f \in Z$. We have $R(a_k)R(x'f) = -R(f)R(x'a_k) - R(x')R(a_kf) + R(a_k)R(x')R(f) + R(f)R(x')R(a_k) + R(x'(a_kf))$.
This implies that $u\cdots R(x_{2r-1})R(a_k)R(x')R(f) \notin I$. Hence, we can assume that $x_{2r} = x$ and for an arbitrary $1 \leq j \leq 2r$, $x_j = x$.

By Lemma 6.20, we have $uP \in (V^\sharp + I) \setminus I$. Hence there exists $v_i, 1 \leq i \leq d$, such that $\mathrm{tr}(uPR(v_i)) \neq 0$.

Now to finish the proof of the lemma it remains to notice that the expression $\text{tr}(uPR(a_{k+1}))$ is skew-symmetric in $a_1, \dots, a_{k+1}$. Lemma is proved. □

REMARK 6.33. If d is even then $I(1) = I(v_i) = (0)$, $1 \leq i \leq d$. Indeed, suppose that $0 \neq u \in I(1)$. For an operator $w = R(x)R(v_{i_1})R(x) \cdots R(x)R(v_{i_{2r}})$, we have $i_1 < d$. Hence $v_{i_1}^2 = 1$ and $uR(x)R(v_{i_1}) = uR(xv_{i_1}) = 0$, because the element u belongs to the eigenvalue 1 with respect to $U(v_{i_1})$. Now let $w = R(v_{i_1})R(x) \cdots R(x)R(v_{i_{2r+1}})$. The element uw has weight $v_{i_1} \cdots v_{i_{2r+1}}$, $r \geq 1$. If d is even then no nonzero element of $\bar{A}$ has this weight. Hence $I(1) = (0)$. We have $I(v_i)v_i \subseteq I(1) = (0)$. Since $I(v_i)$ belongs to the eigenvalue 1 with respect to $U(v_i)$ this implies that $I(v_i) = (0)$.

Consider also the set $\widetilde{W}$ of multiplication operators of the types:

(3) $R(x)R(v_{i_1})R(x) \cdots R(x)R(v_{i_{2r+1}})$, where $1 \leq i_1 < \cdots < i_{2r+1} \leq d$, $r > 1$,

(4) $R(v_{i_1})R(x)R(v_{i_2})R(x) \cdots R(x)R(v_{i_{2r}})$, $r \geq 2$, $1 \leq i_1 < \cdots < i_{2r} \leq d$.

LEMMA 6.34. *For an arbitrary nonzero element $y \in M$ such that $[yA, M] \subseteq I$ there exists an operator $w \in \tilde{W}$ such that $tr(yw) \neq 0$.*

The proof is similar to the proof of the previous lemma.

LEMMA 6.35. *Under the assumption that $I \neq (0)$ we have $dim_F V \geq 2$.*

PROOF. If $dim_F V = 1$ then $V = V_0$ or $V = V_{n/2}$. Let $0 \neq v \in V$. We have $A = I + Z + vZ$.
Hence, $t(IR(M)R(A)R(M)R(A)) = t(IR(M)R(vZ)R(M)R(vZ))$.

Now for arbitrary elements $y \in M$, $f, g \in Z$ we have
$IR(M)R(vf)R(y) = IR(M)R(v)R(yf) \ mod\ I$ by the properties of rescuing operators.

Moreover, from Lemma 1.28 and the Jordan identity applied to the underline parts, we have

$$t(IR(M)\underline{R(vf)R(y)R(vg)}) = t(IR(M)R(vg)R(yf)R(v)) =$$
$$t(IR(M)\underline{R(v)R((yf)g)R(v)}) = (0).$$

This contradicts the results of Chapter 1. Lemma is proved. □

In what follows we will assume that $dim_F V \geq 2$.

Consider the Clifford algebra Cl on $V^\sharp$, that is, the Z-algebra generated by $v_1, \dots, v_d$ with relations $v_i^2 = 1$, $1 \leq i \leq d-1$, $v_d^2 = 1$ if $\dim V_{n/2}$ is even and $v_d^2 = t^n$ if $\dim V_{n/2}$ is odd.

We say that a root element $a \in I$ has weight $v_{i_1} \cdots v_{i_r} \in Cl$, $1 \leq i_1 < \cdots < i_r \leq d$ if for any $1 \leq j \leq d$ the equality $a(U(v_j, v_j^{-1}) - \epsilon)^k = 0$ holds for some k if and only if $(v_{i_1} \cdots v_{i_r})(U(v_j, v_j^{-1}) - \epsilon) = 0$.

If $a \in I$ is a root element and $\operatorname{tr}(aR(x)R(v_{i_1})R(x)\cdots R(x)R(v_{i_{2r}})) \neq 0$ (resp. $\operatorname{tr}(aR(v_{i_1})R(x)R(v_{i_2})R(x)\cdots R(x)R(v_{i_{2r+1}})) \neq 0$) then $aR(x)R(v_{i_1})R(x)\cdots R(x)R(v_{i_{2r}})$ (resp. $aR(v_{i_1})R(x)R(v_{i_2})R(x)\cdots R(x)R(v_{i_{2r+1}})$) belongs to the eigenvalue 1 with respect to each of the operators $U(v_1),\ldots,$ $U(v_{d-1}), U(v_d, v_d^{-1})$.

Hence by Lemma 6.27 a has weight $v_{i_1}\cdots v_{i_{2r}}$ (resp. $v_{i_1}\cdots v_{i_{2r+1}}$).

For $u = v_{i_1}\cdots v_{i_r} \in Cl$ denote $I(u) = \{a \in I \mid a \text{ has weight } u\}$.

LEMMA 6.36. *For an arbitrary operator $w \in W$ and arbitrary elements $a \in I$, $f \in Z$ we have $(af)w - (aw)f \in I$.*

PROOF. Without loss of generality we will assume that a is a root element.

Suppose that $(af)w - (aw)f \notin I$. Then $(af)w \notin I$ or $aw \notin I$.

If a is a root element of J belonging to the eigenvalue 1 with respect to $U(v_j)$, $v_j^2 = 1$, then $aR(x)R(v_j) = 0$.

Consider first the operator $w = R(x)R(v_{i_1})R(x)\cdots R(x)R(v_{i_{2r}})$. The elements $v_{i_1},\ldots,v_{i_{2r-1}}$ are involutions. If a belongs to the eigenvalue 1 with respect to at least one operator $U(v_j)$, $j \in \{i_1,\ldots,i_{2r-1}\}$, then $aw = (af)w = 0$ contrary to our assumption.

Hence a belongs to the eigenvalue -1 with respect to $U(v_{i_1}),\ldots,U(v_{i_{2r-1}})$. This implies $aR(x)R(v_{i_1})R(x)\cdots R(v_{i_{2r-1}}) = aD(x,v_{i_1})\cdots D(x,v_{i_{2r-1}})$. If $i_{2r} = d$ and $v_d^2 = t^n$ then a belongs to the eigenvalue -1 with respect to $U(v_d^{-1}, v_d)$. Otherwise $aw \in I$ and $(af)w \in I$.

We claim that the element $b = aR(x)R(v_{i_1})R(x)\cdots R(x)R(v_{i_{2r-2}})$ lies in I. If $r = 1$ then $a \in I$. Suppose that $r \geq 2$. The element b belongs to the eigenvalue 1 with respect to $U(v_{i_1}),\ldots,U(v_{i_{2r-2}})$ and to the eigenvalue -1 with respect to $U(v_{i_{2r-1}})$. There are no such nonzero elements in A/I. We also notice that $bR(x)R(v_{i_{2r-1}})R(v_{i_{2r}})R(x) \in I$. This easily follows from $R(v_{i_{2r-1}})R(v_{i_{2r}})R(x) = -R(x)R(v_{i_{2r}})R(v_{i_{2r-1}}) + R(v_{i_{2r}}x)R(v_{i_{2r-1}})$ and from Lemma 1.27. So we have proved that
$aR(x)R(v_{i_1})R(x)\cdots R(v_{i_{2r-1}})R(x)R(v_{2r}) = aD(x,v_{i_1})\cdots D(x,v_{i_{2r-1}})D(x,v_{2r}) \bmod I$.

Similarly $(af)w = (af)D(x,v_{i_1})\cdots D(x,v_{i_{2r}})$ *mod* I.

Denote $D(x,v_{i_k}) = D_k$. We have $(af)D_1\cdots D_{2r} = \sum \pm(aD_{j_1}\cdots D_{j_\mu})(fD_{h_1}\cdots D_{h_\nu})$, $\mu + \nu = 2r$. If μ is even and $\nu \neq 0$ then $aD_{j_1}\cdots D_{j_\mu} \in I$. Indeed, this element belongs to the eigenvalue 1 with respect to $U(v_{i_{j_1}}, {v_{i_{j_1}}}^{-1}),\ldots,U(v_{i_{j_\mu}}, {v_{i_{j_\mu}}}^{-1})$ and to the eigenvalue -1 with respect to $U(v_{i_{h_1}}, {v_{i_{h_1}}}^{-1}),\ldots,U(v_{i_{h_\nu}}, {v_{i_{h_\nu}}}^{-1})$. The algebra A/I does not have nonzero elements with this property.

If both μ and ν are odd, then $[aD_{j_1}\cdots D_{j_\mu}, fD_{h_1}\cdots D_{h_\nu}] \in [(MI)A, MV^\sharp] \subseteq I$ by Lemma 6.22.

This implies that $(af)D_1\cdots D_{2r} = (aD_1\cdots D_{2r})f$ *mod* I.

Now let $w = R(v_{i_1})R(x)R(v_{i_2})\cdots R(x)R(v_{i_{2r+1}})$. Then $(af)w = ((av_{i_1})f)w'$, where $w' = R(x)R(v_{i_2})R(x)\cdots R(x)R(v_{i_{2r+1}})$. In view of what we proved above, $((av_{i_1})f)w' = (aR(v_{i_1})w')f = (aw)f$ mod I. Lemma is proved. $\square$

LEMMA 6.37. *For arbitrary integers* $1 \le i < j \le d$ *there exists an element* $a \in I(v_i v_j)$ *such that* $tr(aR(x)R(v_i)R(x)R(v_j)) \ne 0$.

PROOF. Since $I \ne (0)$ it follows that there exist an element $b \in I$ and some numbers $1 \le k < l \le d$ such that $tr(bR(x)R(v_k)R(x)R(v_l)) \ne 0$. If $\{i,j\}$ and $\{k,l\}$ have one element in common, say, $i = k$, $j \ne l$, then
$tr(bD(v_l, v_j)R(x)R(v_k)R(x)R(v_l)) = -tr(bR(x)R(v_k)R(x)R(v_l D(v_l, v_j))) \ne 0$, by Lemma 6.22.

If $\{i,j\} \cap \{k,l\} = \emptyset$ then $tr(bD(v_k, v_i)D(v_l, v_j)R(x)R(v_i)R(x)R(v_j)) \ne 0$. Lemma is proved. □

Combining Lemmas 636 and 6.37 we can find a root element a_{ij} from the completion $\widehat{I}(v_i v_j)$ such that $a_{ij}R(x)R(v_i)R(x)R(v_j) = 1 \ mod \ \widehat{I}$. For every pair $1 \le i < j \le d$ fix such an element.

LEMMA 6.38. *Let* $i_1 < j_1, i_2 < j_2, \ldots, i_k < j_k$ *be 2k distinct integers ranging from 1 to d. Suppose further that* $d \ne 5$. *Then*
$a_{i_1 j_1} R(a_{i_2 j_2}) \cdots R(a_{i_k j_k}) R(x)R(v_{i_1})R(x)R(v_{j_1}) \cdots R(x)R(v_{i_k})R(x)R(v_{j_k}) = 1$ $mod \ \widehat{I}$.

PROOF. As in the proof of Lemma 6.36 we denote $D(x, v_j) = D_j$ and conclude that $a_{i_1 j_1} R(a_{i_2 j_2}) \cdots R(a_{i_k j_k})(R(x)R(v_{i_1})R(x)R(v_{j_1}) \cdots$
$R(x)R(v_{i_k})R(x)R(v_{j_k}) - D_{i_1} D_{j_1} \cdots D_{i_k} D_{j_k}) \in \widehat{I}$.

For $k = 1$ the assertion follows from the choice of $a_{i_1 j_1}$.

Let $a' = a_{i_1 j_1} R(a_{i_2 j_2}) \cdots R(a_{i_{k-1} j_{k-1}})$. By induction we assume that $a' D_{i_1} D_{j_1} \cdots D_{i_{k-1}} D_{j_{k-1}} = 1 \ mod \ \widehat{I}$. Now,

$$(a' a_{i_k j_k}) D_{i_1} D_{j_1} \cdots D_{i_k} D_{j_k} = \sum \pm (a' D_{\mu_1} \cdots D_{\mu_s})(a_{i_k j_k} D_{\nu_1} \cdots D_{\nu_q}),$$

where $\{\mu_1, \ldots, \mu_s, \nu_1, \ldots, \nu_q\} = \{i_1, j_1, i_2, j_2, \ldots, i_k, j_k\}$, $s + q = 2k$.

Suppose that $s < 2(k-1)$. Let s be even. The weight of $a' D_{\mu_1} \cdots D_{\mu_s}$ is $v_{i_1} v_{j_1} \cdots v_{i_{k-1}} v_{j_{k-1}} v_{\mu_1} \cdots v_{\mu_s}$, an even element of the Clifford algebra which does not lie in Z because $s < 2(k-1)$. If $a' D_{\mu_1} \cdots D_{\mu_s} \notin \widehat{I}$ then $s = 2$, $\{\mu_1, \mu_2\} = \{i_k, j_k\}$ and $d = 2k + 1$. However, this is impossible because $a' \in \widehat{I}^2$ for $d > 3$ and therefore $a' D_{\mu_1} D_{\mu_2} \in \widehat{I}$ by Lemma 1.41. Hence, $a' D_{\mu_1} \cdots D_{\mu_s} \in \widehat{I}$ and therefore $(a' D_{\mu_1} \cdots D_{\mu_s})(a_{i_k j_k} D_{\nu_1} \cdots D_{\nu_q}) \in \widehat{I}$.

If s is odd then for the same reason as above we have $a' D_{\mu_1} \cdots D_{\mu_s} \in \widehat{M_I}$. Hence the element $(a' D_{\mu_1} \cdots D_{\mu_s})(a_{i_k j_k} D_{\nu_1} \cdots D_{\nu_q})$ lies in $[\hat{M}_I, \hat{M}]$ and therefore has zero trace. This implies

$$(a' D_{\mu_1} \cdots D_{\mu_s})(a_{i_k j_k} D_{\nu_1} \cdots D_{\nu_q}) \in \widehat{I}.$$

Hence if a summand does not lie in $\widehat{I}$ then $s \ge 2(k-1)$. If $q \le 1$ then the element $a_{i_k j_k} D_{\nu_1} \cdots D_{\nu_q}$ lies either in $\widehat{I}$ or in $\hat{M}_I$ and we can repeat the argument.

Now let $q = 2$, $s = 2(k-1)$. The element $a_{i_k j_k} D_{\nu_1} D_{\nu_2}$ has weight $v_{i_k} v_{j_k} v_{\nu_1} v_{\nu_2}$.

If this element does not lie in $\widehat{I}$ then either $\{\nu_1, \nu_2\} = \{i_k, j_k\}$ or $d = 5$ and $\{\nu_1, \nu_2\} \cap \{i_k, j_k\} = \emptyset$. The latter case though is ruled out by the assumption of

the lemma. Hence $\{\nu_1, \nu_2\} = \{i_k, j_k\}$, $\{\mu_1, \dots, \mu_{2(k-1)}\} = \{i_1, j_1, \dots, i_{k-1}, j_{k-1}\}$. This implies that
$(a' a_{i_k j_k}) D_{i_1} D_{j_1} \cdots D_{i_k} D_{j_k} = (a' D_{i_1} D_{j_1} \cdots D_{i_{k-1}} D_{j_{k-1}})(a_{i_k j_k} D_{i_k} D_{j_k}) = 1 \ mod\ \widehat{I}$.
Lemma is proved. □

LEMMA 6.39. *There exists an element $a \in I(v_1 v_2)$ such that the only operator $w \in W$ with the property $tr(aw) \neq 0$ is $R(x)R(v_1)R(x)R(v_2)$.*

PROOF. If d = 2 then W consists of one element.

If $d = 3$ then W consists of four operators $R(x)R(v_1)R(x)R(v_2)$, $R(x)R(v_1)R(x)R(v_3)$, $R(x)R(v_2)R(x)R(v_3)$, $R(v_1)R(x)R(v_2)R(x)R(v_3)$, the elements $v_1 v_2$, $v_1 v_3$, $v_2 v_3$, $v_1 v_2 v_3$ belong to distinct roots with respect to the action of G. Therefore $I = I(v_1 v_2) + I(v_1 v_3) + I(v_2 v_3) + I(v_1 v_2 v_3)$ is a direct sum of subspaces. For any nonzero element $a \in I(v_1 v_2)$ we have
$tr(aR(x)R(v_1)R(x)R(v_2)) \neq 0$ and $w = R(x)R(v_1)R(x)R(v_2)$ is the only operator of W with this property.

Assume that $d \geq 4$. If d is even, then all products $v_{i_1} \cdots v_{i_k}$, $i_1 < \cdots < i_k$ in Cl belong to distinct roots and therefore for any nonzero element a in $N(v_1 v_2)$ the trace $tr(aW)$ is nonzero only for $w = R(x)R(v_1)R(x)R(v_2)$.

Suppose that d is odd. Then $v_3 \cdots v_d$ and $v_1 v_2$ belong to the same root with respect to the action of G.
Denote $w = R(x)R(v_1)R(x)R(v_2)$, $w' = R(v_3)R(x)R(v_4) \cdots R(x)R(v_d)$.
Suppose that the assertion of the Lemma is wrong. This means that
$0 \neq a \in I(v_1 v_2)$, $tr(aw) \neq 0$ implies that $tr(aw') \neq 0$ (*)

Claim 1 For arbitrary nonzero elements $a, b \in I(v_1 v_2)$ there exist $0 \neq f \in Z$, $0 \neq g \in Z$ such that $af = bg$.
Indeed, if $tr(aw') = 0$ (resp. $tr(bw') = 0$) then by Lemma 6.34 we have $tr(aw) \neq 0$ (resp. $tr(bw) \neq 0$) which contradicts (*). Denote $tr(aw') = g$, $tr(bw') = f$. Then $aw'R(f) - bw'R(g) \in I$. From Lemma 6.36 it follows that $(af - bg)w' \in I$ which together with (*) yields $tr((af - bg)w) = 0$. Hence $af = bg$ by Lemma 6.34.

Remark It is clear that for any i, j, $1 \leq i < j \leq d$ and for any nonzero elements a, $b \in I(v_i v_j)$ there exist $0 \neq f$, $0 \neq g \in Z$ such that $af = bg$.

Claim 2 If $0 \neq a \in I(v_i v_j)$, $1 \leq i < j \leq d$ then

$$tr(aR(x)R(v_i)R(x)R(v_j)) \neq 0.$$

Indeed, if $0 \neq a \in I(v_i v_j)$ is a counterexample to the claim, then, by Lemma 6.36, for an arbitrary $0 \neq f \in Z$ the product af is a counterexample as well. From the claim 1 it follows that for the whole space $I(v_i v_j)$ we have $I(v_i v_j)R(x)R(v_i)R(x) \subseteq I$. This contradicts Lemma 6.37.

Claim 3 Let $1 \leq i, j, k \leq d - 1$ be three distinct integers. Then $I(v_i)I(v_j v_k) = I(v_i v_j)I(v_j v_k) = (0)$.
Indeed, let $a \in I(v_i v_j), b \in I(v_j v_k)$. If $ab \neq 0$ then by claim 2
$tr((ab)R(x)R(v_i)R(x)R(v_k)) \neq 0$, whereas $(ab)R(x)R(v_i)R(x) \in I$. Similarly, if $u \in I(v_i)$ and $0 \neq ub \in I(v_i v_j v_k)$ then $0 \neq (ub)R(v_i) \in I(v_j v_k)$.

We have $(ub)R(v_i)R(x)R(v_j)R(x) \in I$ by Lemma 1.41, which contradicts claim 2

Claim 4 If $1 \leq i < j < k \leq d$ then $(xI(v_iv_j))I(v_jv_k) = (0)$.

Choose arbitrary elements $a \in I(v_iv_j), b \in I(v_jv_k)$. First, we will show that $c = xR(a)R(b)R(x) = 0$. Indeed, $c \in I(v_iv_k)$ and $cR(x)R(v_i)R(x) \in I$. From claim 2 it follows that $c = 0$.

Suppose that $xR(a)R(b) \neq 0$. From Lemma 6.34 it follows that there exists an operator $w \in \tilde{W}$ such that $tr(xR(a)R(b)w) \neq 0$. Since $c = 0$ the operator w should be of the type

$$w = R(v_{i_1})R(x)R(v_{i_2})R(x) \cdots R(v_{i_{2s-1}})R(x)R(v_{i_{2s}}),$$

$1 \leq i_1 < \cdots < i_{2s} \leq d$.

The element v_{i_1} is an involution. If $i_1 \neq i$, $i_1 \neq k$, then $xR(a)R(b)R(v_{i_1}) = 0$. Hence, $\{i_1, \ldots, i_{2s}\} \cap \{1, 2, \ldots, d-1\} \subseteq \{i, k\}$. Hence, $s \leq 1$, a contradiction. The claim is proved.

Suppose that $d = 5$. We showed above that there exists an element $a_{12} \in \widehat{I}(v_1v_2)$ such that $a_{12}R(x)R(v_1)R(x)R(v_2) = 1 \; mod\, \widehat{I}$. Since the element $a_{12}R(x)R(v_1)R(x)$ has weight v_2 it implies that $a_{12}R(x)R(v_1)R(x) = v_2 \; mod \; \widehat{I}$. Hence, for a nonzero element $b_{13} \in I(v_1v_3)$ we have:

$0 \neq v_2b_{13} = a_{12}R(x)R(v_1)R(x)R(b_{13}) \; mod\, \widehat{I}$. We have also $a_{12}R(x)(-R(b_{13})R(x)R(v_1) + R(xb_{13})R(v_1)) = a_{12}R(x)R(xb_{13})R(v_1)$ since $v_1x = v_1b_{13} = 0$ and $a_{12}R(x)R(b_{13}) = 0$ by claim 4.

Furthermore, $R(xb_{13})R(v_1) = R(x)R(b_{13})R(v_1) + R(v_1)R(b_{13})R(x)$.

We have $a_{12}R(x)^2R(b_{13})R(v_1) \in (\widehat{I}(v_1v_2)I(v_1v_3))v_1 = (0)$ by claim 3. Consider the element $a_{12}R(x)R(v_1)R(b_{13}) = xR(a_{12})R(v_1)R(b_{13})$.

If this element is $\neq 0$ then $xR(a_{12})R(v_1)R(b_{13})R(v_4) \neq 0$. Applying the Jordan identity we get $xR(a_{12})\underline{R(v_1)R(b_{13})R(v_4)} =$
$xR(a_{12})(-R(v_4)R(b_{13})R(v_1) + R(b_{13}v_4)R(v_1)) = xR(a_{12})R(b_{13}v_4)R(v_1)$ since $(xa_{12})v_4 = 0$. Now it remains to notice that $b_{13}v_4 \in I(v_1v_3v_4) = I(v_2v_5)$ and therefore $xR(a_{12})R(b_{13}v_4) = 0$, by claim 4, the contradiction.

Now suppose that $d \geq 7$. Since $tr(a_{12}R(x)R(v_1)R(x)R(v_2)) = 1$ it follows from the assumption (*) that $a_{12}w' = f(t^n) \bmod \widehat{I}$, $f \neq 0$, where

$$w' = R(v_3)R(x)R(v_4)R(x) \cdots R(x)R(v_d).$$

Lemma 6.38 implies that $a_{45}R(a_{67}) \cdots R(a_{d-1,d})R(v_3)w' = 1 \bmod \widehat{I}$.

Hence, by Lemma 6.36 $a_{45}R(a_{67}) \cdots R(a_{d-1,d})R(v_3)R(f)w' = f \bmod \widehat{I}$.

Now by (*) and claim 2

$$a_{12} - a_{45}R(a_{67}) \cdots R(a_{d-1,d})R(v_3)R(f) = 0.$$

Since $d \geq 7$ it follows that $a_{12} \in \widehat{I}^2\hat{A}$ and therefore $a_{12}R(x)R(v_1)R(x) \in \widehat{I}$, contradicting the choice of a_{12}. Lemma is proved. □

From now on, by a_{12} we will denote a nonzero root element from $I(v_1v_2)$ such that the only operator $w \in W$ with the property $tr(a_{12}w) \neq 0$ is $R(x)R(v_1)R(x)R(v_2)$.

For $3 \leq j \leq d$ let $a_{1j} = a_{12}D(v_2, v_j)$. For $i \neq 1, j \neq 1$, let $a_{ij} = a_{1j}D(v_1, v_i)$. The element a_{i1}, $i \neq 1$ is defined as $a_{i1} = -a_{1i}$.

LEMMA 6.40. *For arbitrary integers $1 \leq i \neq j \leq d$ we have $a_{ij} = -a_{ji}$.*

PROOF. If $i = 1$ or $j = 1$ then the assertion follows from the definition. Let $i \neq 1$, $j \neq 1$. Suppose that $3 \leq i, j \leq d$. Then $a_{ij} = a_{12}D(v_2, v_j)D(v_1, v_i) = a_{12}R(v_j)R(v_2)R(v_i)R(v_1)$.

From the Jordan identity it follows that $R(v_j)R(v_2)R(v_i) = -R(v_i)R(v_2)R(v_j)$.

That's why

$$a_{ij} = a_{12}R(v_j)R(v_2)R(v_i)R(v_1) = -a_{12}R(v_i)R(v_2)R(v_j)R(v_1) = -a_{ji}.$$

Let $j = 2$, $3 \leq i \leq d$. Then $a_{i2} = a_{12}D(v_1, v_i)$ and $a_{2i} = a_{1i}D(v_1, v_2) = a_{12}D(v_2, v_i)D(v_1, v_2) = a_{12}R(v_i)R(v_2)R(v_2)R(v_1) = a_{12}R(v_i)R(v_1) = a_{12}D(v_i, v_1) = -a_{i2}$.

Lemma is proved. □

LEMMA 6.41. *If $j < d$, $j \neq k$, then $a_{ik} = a_{ij}D(v_j, v_k)$.*

PROOF. The proof is a straightforward examination of all possible cases.

Let $i = 1$. If $k = 2$ then we have to verify that $a_{12} = a_{12}D(v_2, v_j)D(v_j, v_2)$. Indeed, $a_{12}D(v_2, v_j) = -a_{12}R(v_j)R(v_2)$. From the Jordan identity

$$2R(v_j)R(v_2)R(v_j) + R(v_2) = 2R(v_jv_2)R(v_2) + R(v_j^2)R(v_2)$$

it follows that $R(v_j)R(v_2)R(v_j) = 0$. Hence,

$$a_{12}D(v_2, v_j)D(v_j, v_2) = a_{12}R(v_j)R(v_2)R(v_2)R(v_j) = a_{12}R(v_j)R(v_j) = a_{12}.$$

If $k \neq 2$, $j = 2$ then there is nothing to prove.

If $k \neq 2$, $j \neq 2$ then we have to check that $a_{12}D(v_2, v_j)D(v_j, v_k) = a_{12}D(v_2, v_k)$ or equivalently, $-a_{12}R(v_k)R(v_2) = a_{12}R(v_j)\underline{R(v_2)R(v_k)R(v_j)}$. To do that it is sufficient to apply the Jordan identity to the underlined part.

Now suppose that $i \neq 1$. Let $k = 1$. We have to check that $a_{i1} = a_{ij}D(v_j, v_1)$ or, equivalently, $-a_{1i} = a_{ij}D(v_j, v_1) = -a_{1i}D(v_1, v_j)D(v_j, v_1)$. The right hand side is equal to $-a_{1i}R(v_j)R(v_1)R(v_1)R(v_j) = -a_{1i}$.

If j = 1 then the assertion is the definition of a_{ik}.

Now suppose that i,j,k are distinct from 1. Then $a_{ik} = a_{1k}D(v_1, v_i)$, $a_{ij} = a_{1j}D(v_1, v_i)$ and we have to check that $a_{1k}D(v_1, v_i) = a_{1j}D(v_1, v_i)D(v_j, v_k)$. But $D(v_1, v_i)D(v_j, v_k) = D(v_j, v_k)D(v_1, v_i)$ and $a_{1k} = a_{1j}D(v_j, v_k)$ in view of what was proved earlier. Lemma is proved. □

LEMMA 6.42. *For an arbitrary element $f \in Z$ and an arbitrary integer i, $1 \leq i \leq d$ we have $x(v_if) \in \widehat{M_I}$.*

PROOF. Choose $1 \leq k \leq d$, $k \neq i$. We showed above that there exists a root element $u \in I$ such that $tr(uR(x)R(v_k)R(x)R(v_i)) \neq 0$. This implies that $u \in I(v_k v_i)$ and therefore $uR(x)R(v_k)R(x) = v_i g \bmod I$, $g \in Z$.

From Lemma 6.36 it follows that $(ug^{-1})fR(x)R(v_k)R(x) = v_i f \bmod \widehat{I}$.

Hence, $x(v_i f) = ((ug^{-1})f)R(x)R(v_k)R(x)^2 \ mod \ x\widehat{I}$, which implies the assertion of the lemma. Lemma is proved. □

LEMMA 6.43. *For arbitrary integers $1 \leq i < j \leq d$ the only operator $w \in W$ with the property $tr(a_{ij}w) \neq 0$ is $w = R(x)R(v_i)R(x)R(v_j)$.*

PROOF. If d is even then the elements $\{a_{ij}w\}$, $w \in W$ belong to distinct roots with respect to the action of G. That's why we will assume that d is odd and $d \geq 3$.

Suppose at first that $i = 1$, $j \neq 2$. Then $a_{1j} = a_{12}D(v_2, v_j) = -a_{12}R(v_j)R(v_2)$. Let $\mu_1 < \cdots < \mu_{d-2}$ be all integers from 1 to d except 1,j.

Denote $\omega = R(v_{\mu_1})R(x)R(v_{\mu_2}) \cdots R(x)R(v_{\mu_{d-2}})$. Our aim is to prove that $a_{1j}\omega \in \widehat{I}$ since ω is the only operator from W (except $R(x)R(v_1)R(x)R(v_j)$) such that $a_{1j}\omega$ belongs to the root $(1, 1, \ldots, 1)$ with respect to the action of G.

Clearly, $\mu_1 = 2$. Since $d \geq 3$, the element v_2 is an involution. Hence, $a_{12}R(v_j)R(v_2)^2 = a_{12}R(v_j)$. Now we have to prove that
$a_{12}R(v_j)R(x)R(v_{\mu_2})R(x) \cdots R(x)R(v_{\mu_{d-2}}) \in \widehat{I}$.

If $j \neq d$ then $\mu_{d-2} = d$. The expression $R(v_j)R(x)R(v_{\mu_2}) \cdots R(x)R(v_d)$ is skew-symmetric with respect to involutions $v_j, v_{\mu_2}, \ldots$, therefore it lies in $\pm W$ and we can use the property of the element a_{12}.

Let $j = d$. Then $a_{12}R(v_d)R(x)R(v_3) \cdots R(x)R(v_{d-1}) =$
$(a_{12}v_d)D(x, v_3) \cdots D(x, v_{d-1}) = \sum \pm (a_{12}D_{\xi_1} \cdots D_{\xi_s})(v_d D_{\nu_1} \cdots D_{\nu_q})$,
$s + q = d - 3$, $D_{\xi_1}, \ldots, D_{\xi_s}, D_{\nu_1}, \ldots, D_{\nu_q}$ is a permutation of $D(x, v_3), \ldots,$
$D(x, v_{d-1})$; $\xi_1 < \cdots < \xi_s$; $\nu_1 < \cdots < \nu_q$.

If s is even then $a_{12}D_{\xi_1} \cdots D_{\xi_s} \in \widehat{I}$ because of the assumption about a_{12}. If s is odd, then $a_{12}D_{\xi_1} \cdots D_{\xi_s} \in \widehat{M_I}$ and therefore
$tr([a_{12}D_{\xi_1} \cdots D_{\xi_s}, v_d D_{\nu_1} \cdots D_{\nu_q}]) = 0$.

We have finished the case $i = 1$.

Now let $1 < i < j$.

Then $a_{ij} = a_{1j}D(v_1, v_i) = -a_{1j}R(v_i)R(v_1)$. Let $1 = \mu_1 < \mu_2 < \cdots < \mu_{d-2}$ be all numbers from 1 to d except i,j. We will prove that
$a_{ij}R(v_{\mu_1})R(x) \cdots R(x)R(v_{\mu_{d-2}}) = -a_{1j}R(v_i)R(x)R(v_{\mu_2}) \cdots R(x)R(v_{\mu_{d-2}}) \in \widehat{I}$.

If $i < d$ then $R(v_i)R(x)R(v_{\mu_2}) \cdots R(x)R(v_{\mu_{d-2}})$ lies in $\pm W$, hence we can apply what was proved about a_{1j}. If $i = d$ then

$$a_{1j}R(v_d)R(x)R(v_{\mu_2}) \cdots R(x)R(v_{\mu_{d-2}}) = (a_{1j}v_d)D(x, v_{\mu_2}) \cdots D(x, v_{\mu_{d-2}})$$

and we need only to repeat the arguments above. Lemma is proved. □

LEMMA 6.44. *(1) For any integers $1 \leq i \neq j \leq d$ we have $a_{ij}^2 = 0$,*

(2) for arbitrary distinct integers $1 \le i, j, k \le d$ we have $a_{ij}a_{ik} = 0$.

PROOF. To prove (1) it is sufficient to check that for an arbitrary operator $w \in W$ we have $tr(a_{ij}^2 w) = 0$. Considering roots with respect to the action of G we see that this is the case if d is even. If d is odd then the only candidate for a counterexample is $w = R(v_1)R(x)R(v_2)R(x)\cdots R(x)R(v_d)$.

As we have already noticed many times:
$a_{ij}^2 w = a_{ij}^2 D(v_1, x)\cdots D(v_{d-1}, x)R(v_d)$;
$a_{ij}^2 D(v_1, x)\cdots D(v_{d-1}, x) = \sum \pm (a_{ij}D_{\xi_1}\cdots D_{\xi_s})(a_{ij}D_{\nu_1}\cdots D_{\nu_q})$; $s + q = d - 1$;
$D_{\xi_1}, \dots, D_{\xi_s}, D_{\nu_1}, \dots, D_{\nu_q}$ is a permutation of $D(v_1, x), \dots, D(v_{d-1}, x)$;
$\xi_1 < \cdots < \xi_s; \nu_1 < \cdots < \nu_q$.

If s and q are even then at least one of $a_{ij}D_{\xi_1}\cdots D_{\xi_s}$, $a_{ij}D_{\nu_1}\cdots D_{\nu_q}$ lies in $\widehat{I}$. If s and q are odd, then at least one of the elements
$a_{ij}D_{\xi_1}\cdots D_{\xi_s}$, $a_{ij}D_{\nu_1}\cdots D_{\nu_q}$ lies in $\widehat{M_I}$.

Indeed, the element $a_{ij}D_{\xi_1}\cdots D_{\xi_{s-1}}$ has weight $v_iv_jv_{\xi_1}\cdots v_{\xi_{s-1}}$. If $a_{ij}D_{\xi_1}\cdots D_{\xi_{s-1}} \in Z + \widehat{I}$ then $a_{ij}D_{\xi_1}\cdots D_{\xi_s} \in \widehat{M_I}$ by Lemma 6.42.

An equality $a_{ij}D_{\xi_1}\cdots D_{\xi_{s-1}} = fv_k \ mod\ \widehat{I}$, $0 \ne f \in Z$ is impossible by Lemma 6.43.

Hence $tr([a_{ij}D_{\xi_1}\cdots D_{\xi_s}, a_{ij}D_{\nu_1}\cdots D_{\nu_q}]) = 0$, which implies $[a_{ij}D_{\xi_1}\cdots D_{\xi_s}, a_{ij}D_{\nu_1}\cdots D_{\nu_q}] \in \widehat{M_I}$. The part (1) is proved.

Let us prove the part (2). One of the numbers j,k is distinct from d. Let $j \ne d$. Then, by Lemma 6.41 $a_{ik} = a_{ij}D(v_j, v_k)$ and therefore $a_{ij}(a_{ij}D(v_j, v_k)) = \frac{1}{2}a_{ij}^2 D(v_j, v_k) = 0$. Lemma is proved. □

LEMMA 6.45. *For any distinct integers $1 \le i, j, k, l \le d$ we have* $D(a_{ij}, a_{kl}) = 0$.

PROOF. Suppose that $i < d$. Then $a_{kl} = (a_{kl}v_i)v_i$ and $D(a_{ij}, a_{kl}) = D(a_{ij}, (a_{kl}v_i)v_i) = D(a_{ij}(a_{kl}v_i), v_i)$. At least one of the numbers k, l is $\ne d$. Let $k < d$. By Lemma 6.41 $a_{kl}v_i = -a_{il}v_k$. Since a_{ij} belongs to eigenvalue 1 with respect to $U(v_k)$ we have $D(a_{ij}, v_k) = 0$. Now it follows that $a_{ij}(a_{il}v_k) = (a_{ij}a_{il})v_k = 0$ by Lemma 6.44. Lemma is proved. □

LEMMA 6.46. *Let $1 \le i_1, \dots, i_{2s} \le d$ be $2s$ distinct integers. The expression $a_{i_1i_2}R(a_{i_3i_4})\cdots R(a_{i_{2s-1}i_{2s}})$ is skew-symmetric in $i_1, i_2, \dots, i_{2s}$.*

PROOF. By Lemma 6.45 it is sufficient to prove that $a_{ij}a_{kl}$ is skew-symmetric in i, j, k, l, that is, that $a_{ij}a_{kl} + a_{il}a_{kj} = 0$. Let $j \ne d$. Then $a_{il} = a_{ij}D(v_j, v_l)$ and $a_{il}a_{kj} = a_{ij}D(v_j, v_l)a_{kj} = (a_{ij}a_{kj})D(v_j, v_l) - a_{ij}(a_{kj}D(v_j, v_l)) = -a_{ij}a_{kl}$. Lemma is proved. □

Denote $a_{i_1\cdots i_{2s}} = a_{i_1i_2}R(a_{i_3i_4})\cdots R(a_{i_{2s-1}i_{2s}})$.

LEMMA 6.47. *For any distinct $2s+1$ integers $1 \le i_1, \dots, i_{2s+1} \le d$ the expression*

$$a_{i_1 \cdots i_{2s+1}} = (-1)^{k+1} a_{i_1 \cdots \hat{i_k} \cdots i_{2s}} v_{i_k},$$

where ˆ means that the k-th term has been ommited, is skew-symmetric in $i_1, \dots, i_{2s+1}$.

The proof consists of references to Lemmas 6.41 and 6.46.

Let W_1 be the set of operators of the type

$$R(x)R(v_{i_1})R(x)R(v_{i_2}) \cdots R(x)R(v_{i_{2r}}),$$

where $1 \le i_1, \dots, i_{2r} \le d$ are distinct numbers, or of the type

$$R(v_{i_1})R(x)R(v_{i_2})R(x) \cdots R(x)R(v_{i_{2r+1}}),$$

where again $1 \le i_1, \dots, i_{2r+1} \le d$ are distinct numbers. Unlike in the definition of the set W we do not assume that $i_1 < i_2 < \cdots$

LEMMA 6.48. *(1) Let $1 \le i_1, \dots, i_{2r} \le d$ be distinct integers. For an operator $w \in W_1$ we have $tr(a_{i_1 \cdots i_{2r}} w) \ne 0$ if and only if*

$$w = R(x)R(v_{j_1})R(x) \cdots R(v_{j_{2r}}),$$

where $j_1, \dots, j_{2r}$ is a permutation of $i_1, \dots, i_{2r}$.

(2) Let $1 \le i_1, \dots, i_{2r+1} \le d$ be distinct numbers. For an operator $w \in W_1$ we have $tr(a_{i_1 \cdots i_{2r+1}} w) \ne 0$ if and only if

$$w = R(v_{j_1})R(x)R(v_{j_2}) \cdots R(x)R(v_{j_{2r+1}}),$$

where $j_1, \dots, j_{2r+1}$ is a permutation of $i_1, \dots, i_{2r+1}$.

Remark. In the proof of Lemma 6.48 we will need Lemma 6.38 which was formulated and proved only for $d \ne 5$. However, in view of Lemma 6.38 and the new meaning that we gave to a_{12} and a_{ij} after we proved Lemma 6.39 the difficulty in $d = 5$ disappears. Thus we may assume that Lemma 6.38 is valid for all $d \ge 2$.

PROOF. Lemmas 6.46 and 6.38 imply that if $j_1, \dots, j_{2r}$ is a permutation of $i_1, \dots, i_{2r}$ then $tr(a_{i_1, \dots, i_{2r}} R(x)R(v_{j_1})R(x) \cdots R(x)R(v_{j_{2r}})) \ne 0$.
Our aim is to prove the "only if" part.

Let $w = R(x)R(v_{j_1})R(x) \cdots R(x)R(v_{j_{2r}}) \in W_1$ and $tr(a_{i_1, \dots, i_{2r}} w) \ne 0$. The element $a_{i_1, \dots, i_{2r}} w$ has weight $v_{i_1} \cdots v_{i_{2r}} v_{j_1} \cdots v_{j_{2r}}$. It can have nonzero trace only if $\{i_1, \dots, i_{2r}\} = \{j_1, \dots, j_{2r}\}$ or d is odd and $v_{i_1} \cdots v_{i_{2r}} v_{j_1} \cdots v_{j_{2r}} = \pm v_1 \cdots v_d$, the latter being impossible because d is odd.

Let $w = R(v_{j_1})R(x)R(v_{j_2}) \cdots R(x)R(v_{j_{2r+1}})$. Then $tr(a_{i_1, \dots, i_{2r}} w) \ne 0$ implies $v_{i_1} \cdots v_{i_{2r}} v_{j_1} \cdots v_{j_{2r+1}} = \pm v_1 \cdots v_d$, so $\{j_1, \dots, j_{2r+1}\} = \{1, 2, \dots, d\} - \{i_1, \dots, i_{2r}\}$.

If a is an arbitrary root element of J and $v_j^2 = 1$ then $aR(x)R(v_j) = 0$ or $aR(v_j)R(x) = 0$.
In the first case $aR(v_j)R(x) = -aD(x, v_j)$, in the second case $aR(x)R(v_j) = aD(x, v_j)$.
If $d \in \{j_1, \dots, j_{2r+1}\}$ let $j_k = d$. If $d \notin \{j_1, \dots, j_{2r+1}\}$ let $k = 2r + 1$.

In both cases $a_{i_1\cdots i_{2r}}w = \pm a_{i_1\cdots i_{2r}}D_{j_1}\cdots D_{j_{k-1}}R(v_d)D_{j_{k+1}}\cdots D_{j_{2r+1}}$, where $D_j = D(x, v_j)$. Furthermore,

$$R(v_d)D_{j_{k+1}}\cdots D_{j_{2r+1}} \in \sum D_{\mu_1}\cdots D_{\mu_l}R(A) + \sum D_{\nu_1}\cdots D_{\nu_q}R(M);$$

$\{\mu_1, \dots, \mu_l\}$ and $\{\nu_1, \dots, \nu_q\}$ are subsets of $\{j_{k+1}, \dots, j_{2r+1}\}$; $k-1+l$ is even, $k-1+q$ is odd.

Now it is sufficient to show that if $\lambda_1, \dots, \lambda_{2q}$ are distinct numbers from $\{1, 2, \dots, d-1\} - \{i_1, \dots, i_{2r}\}$ then $a_{i_1\cdots i_{2r}}D_{\lambda_1}\cdots D_{\lambda_{2q}} \in \widehat{I}$.

Denote $a' = a_{i_1\cdots i_{2r-2}}$, so $a = a'a_{i_{2r-1}i_{2r}}$. We have

$$aD_{\lambda_1}\cdots D_{\lambda_{2q}} = \sum \pm(a'D_{\xi_1}\cdots D_{\xi_s})(a_{i_{2r-1}i_{2r}}D_{\eta_1}\cdots D_{\eta_p})$$

where $\xi_1, \dots, \xi_s, \eta_1, \dots, \eta_p$ is a permutation of $\lambda_1, \dots, \lambda_{2q}, s+p = 2q$.

Let s and p be even. Then

$$a_{i_{2r-1}i_{2r}}D_{\eta_1}\cdots D_{\eta_p} = a_{i_{2r-1}i_{2r}}R(v_{\eta_1})R(x)R(v_{\eta_2})R(x)\cdots R(v_{\eta_p})R(x).$$

remark that since $\eta_1, \dots, \eta_p < d$ we have

$$R(v_{\eta_1})R(x)R(v_{\eta_2})R(x)\cdots R(v_{\eta_p})R(x) \in \pm W.$$

By Lemma 6.38 $tr(a_{i_{2r-1}i_{2r}}R(v_{\eta_1})R(x)R(v_{\eta_2})R(x)\cdots R(v_{\eta_p})R(x)) = 0$. Hence,

$$a_{i_{2r-1}i_{2r}}R(v_{\eta_1})R(x)R(v_{\eta_2})R(x)\cdots R(v_{\eta_p})R(x) = v_jg \ mod\ \widehat{I},$$

where g is a Laurent series in t^n.

Again $R(v_{\eta_1})R(x)\cdots R(v_{\eta_p})R(x)R(v_j) \in \pm W$ and therefore $tr(v_j^2g) = 0$. This implies $g = 0$.

Now let s and p be odd. Then arguing as above we conclude that $a_{i_{2r-1}i_{2r}}D_{\eta_1}\cdots D_{\eta_{p-1}} \in \widehat{I}$ and $a'D_{\xi_1}\cdots D_{\xi_{s-1}} \in \widehat{I}$.

Hence $(a'D_{\xi_1}\cdots D_{\xi_s})(a_{i_{2r-1}i_{2r}}D_{\eta_1}\cdots D_{\eta_p}) \in [\widehat{M_I}, \widehat{M_I}] \subseteq \widehat{I}$. The assertion (1) is proved.

(2) First we will prove the if part. Let $1 \le i_1, \dots, i_{2r+1} \le d$ be distinct numbers. Let $j_1, \dots, j_{2r+1}$ be a permutation of $i_1, \dots, i_{2r+1}$. Let $a' = a_{i_1\cdots\hat{j_1}\cdots i_{2r+1}}$, $a = \pm a'v_{j_1}$. If $v_{j_1}^2 = 1$ then $(a'v_{j_1})R(v_{j_1}) = a'$, so $tr(a'v_{j_1})R(v_{j_1})R(x)\cdots R(x)R(v_{j_{2r+1}}) = tr(a'R(x)R(v_{j_2}\cdots R(x)R(v_{j_{2r+1}})) \neq 0$ by Lemma 6.43 and by the remark above.

Let $j_1 = d$ and $v_d^2 = t^n$.

Then $tr(aR(v_{j_1})R(x)\cdots R(v_{j_{2r+1}})) = \pm tr(a'R(v_d)R(v_d)D_{j_2}\cdots D_{j_{2r+1}})$.

We have $a'R(v_d)R(v_d)D_{j_2}\cdots D_{j_{2r+1}} =$
$\sum \pm(a'D_{\lambda_1}\cdots D_{\lambda_k})R(v_dD_{\mu_1}\cdots D_{\mu_s})R(v_dD_{\nu_1}\cdots D_{\nu_q})$,
$k+s+q = 2r$.

Suppose that $q \neq 0$. If q is even then $v_dD_{\nu_1}\cdots D_{\nu_q}$ has weight $v_dv_{\nu_1}\cdots v_{\nu_q}$. If $v_dD_{\nu_1}\cdots D_{\nu_q} \notin \widehat{I}$ then d is odd and $\{\nu_1, \dots, \nu_q\} = \{1, 2, \dots, d-1\}$, so $k = s = 0$. In this case

$$(a'D_{\lambda_1}\cdots D_{\lambda_k})R(v_dD_{\mu_1}\cdots D_{\mu_s})R(v_dD_{\nu_1}\cdots D_{\nu_q}) = a'R(v_d)R(v_dD_{\nu_1}\cdots D_{\nu_q}) \in \widehat{I}.$$

If q is odd and $q \geq 3$ then $v_d D_{\nu_1} \cdots D_{\nu_{q-1}} \in \widehat{I}$, so the element $(a' D_{\lambda_1} \cdots D_{\lambda_k}) R(v_d D_{\mu_1} \cdots D_{\mu_s}) R(v_d D_{\nu_1} \cdots D_{\nu_q})$ lies in $[\widehat{M}, \widehat{M_I}]$ and has zero trace. Finally, if q = 1 then $v_d D_{\nu_1} \in \widehat{M_I}$ by Lemma 6.42.

So, in a summand with a nonzero trace q is equal to zero.

Suppose that $s \neq 0$. If s is even then so is k and therefore

$$a' D_{\lambda_1} \cdots D_{\lambda_k} R(v_d D_{\mu_1} \cdots D_{\mu_s}) \in [\widehat{M_I}, \widehat{M_I}] \subseteq \widehat{I}.$$

Hence, $tr(aR(v_d) D_{j_2} \cdots D_{j_{2r+1}}) = \pm tr((a' D_{j_2} \cdots D_{j_{2r+1}}) R(v_d) R(v_d)) = tr(a' D_{j_2} \cdots D_{j_{2r+1}}) t^n \neq 0$.

Now let us prove the "only if" part. Let $w \in W_1$ and $tr(a_{i_1 \cdots i_{2r+1}} w) \neq 0$.

Now let $w = R(x) R(v_{j_1}) R(x) \cdots R(x) R(v_{j_{2s}})$. Since the element $a_{i_1 \cdots i_{2r+1}} w$ has weight $v_{i_1} \cdots v_{i_{2r+1}} v_{j_1} \cdots v_{j_{2s}}$ it follows that d is odd and $\{j_1, \dots, j_{2s}\} = \{1, 2, \dots, d\} - \{i_1, \dots, i_{2r+1}\}$.

Then $a_{i_1, \dots, i_{2r+1}} = a_{i_1, \dots, i_{2r}} v_{i_{2r+1}}$ and $a_{i_1, \dots, i_{2r+1}} w = a_{i_1, \dots, i_{2r}} w'$, where $w' = R(v_{i_{2r+1}}) w \in W_1$, which contradicts (1).

Let $w = R(v_{j_1}) R(x) \cdots R(x) R(v_{j_{2s+1}})$. The element $a_{i_1, \dots, i_{2r+1}} w$ has weight $v_{i_1} \cdots v_{i_{2r+1}} v_{j_1} \cdots v_{j_{2s+1}}$. Since $\mathrm{tr}(a_{i_1, \dots, i_{2r+1}} w) \neq 0$, it follows that $\{j_1, \dots, j_{2s+1}\} = \{i_1, \dots, i_{2r+1}\}$. Lemma is proved. □

For an ordered subset $\Pi = \{i_1, \dots, i_k\}$ of $\{1, 2, \dots, d\}$ consisting of ≥ 2 elements, denote $a_\pi = a_{i_1 \cdots i_k}$.

LEMMA 6.49. *$\widehat{I} = \sum a_\pi \widehat{Z}$, where π runs over all ordered subsets $\{i_1 < \cdots < i_k\}$ of $\{1, \dots, d\}$ containing at least two elements.*

PROOF. Let $0 \neq a \in \widehat{I}$ be a root element of the weight $v_{i_1} \cdots v_{i_k} \in Cl$, $i_1 < \cdots < i_k$. If d is even then $k \geq 2$ since $\widehat{I}(1) = \widehat{I}(v_i) = (0)$. In this case there exists only one operator $w \in W$ such that $tr(\widehat{I}(v_{i_1} \cdots v_{i_k}) w) \neq (0)$. Let $tr(aw) = f$, $0 \neq f \in \widehat{Z}$. By Lemma 6.48 $tr(a_{i_1 \cdots i_k} w) = g \neq 0$. Now $tr((a - a_{i_1 \cdots i_k}(g^{-1} f)) w) = 0$, which implies that $a = a_{i_1 \cdots i_k}(g^{-1} f)$.

Now suppose that d is odd. If $2 \leq k \leq d-2$ then there are two operators $w, w' \in W$ such that

$$tr(I(v_{i_1} \cdots v_{i_k}) w) \neq (0), \; tr(I(v_{i_1} \cdots v_{i_k}) w') \neq (0).$$

These two operators correspond to a disjoint union $\{1, 2, \dots, d\} = \pi \cup \tau$, $|\pi| \geq 2$, $|\tau| \geq 2$, $tr(a_\pi w) = g_\pi \neq 0$, $tr(a_\tau w') = g_\tau \neq 0$. Let $tr(aw) = f$, $tr(aw') = h$. Then $tr((a - a_\pi(f g_\pi^{-1}) - a_\tau(h g_\tau^{-1})) w) = tr((a - a_\pi(f g_\pi^{-1}) - a_\tau(h g_\tau^{-1})) w') = 0$ which implies that $a = a_\pi(f g_\pi^{-1}) + a_\tau(h g_\tau^{-1})$.

If $k = 0$ or d then arguing as above we get $a \in a_{12 \cdots d} \hat{Z}$.

Let $k = 1$ or, equivalently, $d - 1$, the element a has weight v_i. Let $\pi = \{1, 2, \dots, d\} \setminus \{i\}$. Then $a \in a_\pi \hat{Z}$. Lemma is proved. □

LEMMA 6.50. *Let π, τ be two nonempty subsets of $\{1, 2, \dots, d\}$ of even orders. Suppose that $\pi \cap \tau \neq \emptyset$. Then for arbitrary elements $f, g \in Z$ we have $(a_\pi f)(a_\tau g) = 0$.*

PROOF. If $a = (a_\pi f)(a_\tau g) \neq 0$ then there exists an operator $w = D(x, v_{i_1}) \cdots D(x, v_{i_{2r}})$, $i_1 < \cdots < i_{2r}$ or $w = D(x, v_{i_1}) \cdots D(x, v_{i_{2r}}) R(v_{i_{2r+1}})$, $i_1 < \cdots < i_{2r+1}$, such that $tr(aw) \neq 0$.

We have $(a_\pi f)(a_\tau g) D(x, v_{i_1}) \cdots D(x, v_{i_{2r}}) =$ $\sum \pm ((a_\pi f) D_{\mu_1} \cdots D_{\mu_s})((a_\tau g) D_{\nu_1} \cdots D_{\nu_q})$, where $D_j = D(x, v_j)$; $\mu_1 < \cdots < \mu_s$; $\nu_1 < \cdots < \nu_q$; $\mu_1, \ldots, \mu_s, \nu_1, \ldots, \nu_q$ is a permutation of $i_1, \ldots, i_{2r}$.

Suppose that $((a_\pi f) D_{\mu_1} \cdots D_{\mu_s})((a_\tau g) D_{\nu_1} \cdots D_{\nu_q}) \notin \widehat{I}$. Let s, q be even. The element $(a_\pi f) D_{\mu_1} \cdots D_{\mu_s}$ can lie outside of $\widehat{I}$ in two cases:

(1) when $\{\mu_1, \ldots, \mu_s\} = \pi$, and

(2) when $\{\mu_1, \ldots, \mu_s\} \cap \pi = \emptyset$, d is odd and $\{\mu_1, \ldots, \mu_s\} \cup \pi = \{1, 2, \ldots, d\} - \{j\}$.

In the second case $(a_\pi f) D_{\mu_1} \cdots D_{\mu_s} R(v_j) =$ $(a_\pi f) R(v_{\mu_1}) R(x) R(v_{\mu_2}) R(x) \cdots R(v_{\mu_s}) R(x) R(v_j) \bmod \widehat{I}$. This contradicts Lemma 6.48. Hence $\{\mu_1, \ldots, \mu_s\} = \pi$. Similarly $\{\nu_1, \ldots, \nu_q\} = \tau$. But $\{\mu_1, \ldots, \mu_s\} \cap \{\nu_1, \ldots, \nu_q\} = \emptyset$, whereas $\pi \cap \tau \neq \emptyset$.

Now suppose that both s and q are odd. If $(a_\pi f) D_{\mu_1} \cdots D_{\mu_s} \notin \widehat{M_I}$ then $(a_\pi f) D_{\mu_1} \cdots D_{\mu_{s-1}} \notin \widehat{I}$. We have shown that this is possible only if $\pi = \{\mu_1, \ldots, \mu_{s-1}\}$. Then $(a_\pi f) D_{\mu_1} \cdots D_{\mu_{s-1}} \in \widehat{Z} + \widehat{I}$, which implies $(a_\pi f) D_{\mu_1} \cdots D_{\mu_s} \in (\widehat{Z} + \widehat{I}) D_{\mu_s} \subseteq \widehat{M_I}$.

Similarly $(a_\tau g) D_{\nu_1} \cdots D_{\nu_q} \in \widehat{M_I}$. By Lemma 1.41 $((a_\pi f) D_{\mu_1} \cdots D_{\mu_s})((a_\tau g) D_{\nu_1} \cdots D_{\nu_q}) \in \widehat{I}$. Lemma is proved. □

LEMMA 6.51. *For arbitrary integers $1 \leq i_1 < \ldots < i_{2r} \leq d$, an arbitrary $f \in Z$ we have $a_{i_1 \cdots i_{2r}}(v_{i_1} f) = 0$.*

PROOF. Let $j_1, \ldots, j_{d-2r+1}$ be the complement of $\{i_2, \ldots, i_{2r}\}$ in the set $\{1, 2, \ldots, d\}$.

Let $w = R(v_{i_2}) R(x) R(v_{i_3}) \cdots R(x) R(v_{i_{2r}})$, if $r \geq 2$. If d is odd, then we consider also the operator $w' = R(x) R(v_{j_1}) R(x) \cdots R(x) R(v_{j_{d-2r+1}})$.

To prove that $a_{i_1 \cdots i_{2r}}(v_{i_1} f) = 0$ it is sufficient to verify that $tr(a_{i_1 \cdots i_{2r}} R(v_{i_1} f) w) = 0$ if $r \geq 2$, and $tr(a_{i_1 \cdots i_{2r}} R(v_{i_1} f) w') = 0$ if d is odd.

We have $a_{i_1 \cdots i_{2r}} R(v_{i_1} f) w = a_{i_1 \cdots i_{2r}} R(v_{i_1} f) D_{i_2} \cdots D_{i_{2r-1}} R(v_{i_{2r}})$; $a_{i_1 \cdots i_{2r}} R(v_{i_1} f) D_{i_2} \cdots D_{i_{2r-1}} = \sum (a_{i_1 \cdots i_{2r}} D_{\mu_1} \cdots D_{\mu_s})(v_{i_1} f D_{\nu_1} \cdots D_{\nu_q})$, where $\mu_1, \ldots, \mu_s, \nu_1, \ldots, \nu_q$ is a permutation of $i_2, \ldots, i_{2r-1}$.

For every summand the element $a_{i_1 \cdots i_{2r}} D_{\mu_1} \cdots D_{\mu_s}$ lies either in $\widehat{I}$ or in $\widehat{M_I}$ depending on whether s is even or odd.

Moreover, if s and q are odd then $(v_{i_1} f) D_{\nu_1} \cdots D_{\nu_q} \in \widehat{M_I}$. Indeed, for $q > 1$ $(v_{i_1} f) D_{\nu_1} \cdots D_{\nu_{q-1}}$ has weight $v_{i_1} v_{\nu_1} \cdots v_{\nu_{q-1}}$ and thus lies in $\widehat{I}$.

Hence, $(a_{i_1 \cdots i_{2r}} D_{\mu_1} \cdots D_{\mu_s})((v_{i_1} f) D_{\nu_1} \cdots D_{\nu_q}) \in \widehat{I}$ by Lemma 1.41.

Now, suppose that d is odd. We have $a_{i_1\cdots i_{2r}}R(v_{i_1}f)w' =$ $a_{i_1\cdots i_{2r}}R(v_{i_1}f)D_{j_1}\cdots D_{j_{d-2r+1}} \bmod \widehat{I}$; and as above $a_{i_1\cdots i_{2r}}R(v_{i_1}f)D_{j_1}\cdots D_{j_{d-2r+1}} = \sum \pm(a_{i_1\cdots i_{2r}}D_{\mu_1}\cdots D_{\mu_s})((v_{i_1}f)D_{\nu_1}\cdots D_{\nu_q})$, where $\mu_1,\dots,\mu_s,\nu_1,\dots,\nu_q$ is a permutation of $j_1,\dots,j_{d-2r+1}$.

Suppose that $tr((a_{i_1\cdots i_{2r}}D_{\mu_1}\cdots D_{\mu_s})((v_{i_1}f)D_{\nu_1}\cdots D_{\nu_q})) \neq 0$.
Let s,q be even. Since d is odd and $|\{i_1,\dots,i_{2r}\}\cap\{\mu_1,\dots,\mu_s\}| \le 1$ it follows that $a_{i_1\cdots i_{2r}}D_{\mu_1}\cdots D_{\mu_s}$ can not have weight 1. Suppose that $a_{i_1\cdots i_{2r}}D_{\mu_1}\cdots D_{\mu_s}$ has weight v_k. This is only possible if $\{i_1,\dots,i_{2r}\}\cap\{\mu_1,\dots,\mu_s\} = \emptyset$ and $\{i_1,i_2,\dots,i_{2r},\mu_1,\dots,\mu_s\} = \{1,2,\dots,d\} - \{k\}$. Then $tr(a_{i_1\cdots i_{2r}}D_{\mu_1}\cdots D_{\mu_s})R(v_k)) =$ $tr(a_{i_1\cdots i_{2r}}R(v_{\mu_1})R(x)R(v_{\mu_2})R(x)\cdots R(v_{\mu_s})R(x)R(v_k)) \neq 0$, which contradicts Lemma 6.48.

Let s,q be odd. Then $a_{i_1\cdots i_{2r}}D_{\mu_1}\cdots D_{\mu_{s-1}} \in \widehat{I}$ and therefore $a_{i_1\cdots i_{2r}}D_{\mu_1}\cdots D_{\mu_s} \in \hat{M}_I$. This contradicts Lemma 1.28. Lemma is proved. □

LEMMA 6.52. *For arbitrary elements $f,g \in Z$, arbitrary distinct $1 \le i_1 < \cdots < i_{2r} \le d$ we have $D(a_{i_1\cdots i_{2r}}f, g) = 0$.*

PROOF. Since $i_1 < d$ it follows that $v_{i_1}^2 = 1$ and therefore $g = (gv_{i_1})v_{i_1}$. We have $D(a_{i_1\cdots i_{2r}}f,(gv_{i_1})v_{i_1}) = D((a_{i_1\cdots i_{2r}}f)(gv_{i_1}),v_{i_1}) + D((a_{i_1\cdots i_{2r}}f)v_{i_1},gv_{i_1})$; $(a_{i_1\cdots i_{2r}}f)(gv_{i_1}) = (a_{i_1\cdots i_{2r}}(gv_{i_1}))f = 0$ by Lemma 6.51 and similarly $(a_{i_1\cdots i_{2r}}f)v_{i_1} = 0$. Lemma is proved. □

LEMMA 6.53. *(1) For arbitrary distinct elements $1 \le i \neq j \le d$ we have $v_i a_{ij} = 0$.*

(2) $AR(a_{ij})^2 = 0$.

PROOF. If d is even then (1) follows from the remark after Lemma 6.20. If $v_i^2 = 1$ then the assertion (1) is obvious. Suppose therefore that $i = d$, $v_d^2 = t^n$ and so $d \ge 3$. There exists a number $1 \le k \le d-1$ such that $k \neq j$. By Lemma 6.41 we have $a_{dj} = a_{kj}D(v_k,v_d)$. Since $a_{dj}R(v_d) = -a_{kj}R(v_d)R(v_k)R(v_d) = \frac{-1}{2}a_{kj}(-R(t^n v_k) + R(t^n)R(v_k)) = 0$ by Lemma 6.52.

Let's prove the assertion (2). If $f \in Z$ then by Lemmas 6.52 and 6.44 (1) we have $(fa_{ij})a_{ij} = f(a_{ij}^2) = 0$. Again by Lemma 6.52 and by the part (1) $(v_i f)a_{ij} = (v_i a_{ij})f = 0$. Similarly $(v_j f)a_{ij} = 0$. Let $k \neq i, k \neq j$. As above $(v_k f)R(a_{ij})^2 = v_k R(a_{ij})^2 R(f)$. If $v_k^2 = 1$ then $D(v_k, a_{ij}) = 0$ and $v_k R(a_{ij})^2 = v_k(a_{ij}^2) = 0$.

Suppose that $k = d$, $v_d^2 = t^n$. The element $v_d R(a_{ij})^2$ belongs to the eigenvalue 1 with respect to $U(v_d, v_d^{-1})$. That's why if $v_d R(a_{ij})^2 \neq 0$ then $2v_d R(a_{ij})^2 R(v_d) = 2a_{ij}R(v_d)R(a_{ij})R(v_d) = a_{ij}(-R(a_{ij}v_d^2) + R(a_{ij})R(t^n) + 2R(v_d)R(a_{ij}v_d)) \neq 0$.

It remains to show that $a_{ijd}^2 = 0$. We have $a_{ijd} = a_{jd}v_i$. Hence $(a_{jd}v_i)R(a_{jd}v_i) = a_{jd}^2 = 0$.
Now we have to show that $(a_\pi f)R(a_{ij})^2 = 0$ for an ordered subset $\pi = \{i_1,\dots,i_k\}$ of $\{1,2,\dots,d\}$, $|\pi| \ge 2$.

Let $|\pi|$ be even. If $\pi \cap \{i,j\} \neq \emptyset$ then $(a_\pi f)a_{ij} = 0$ by Lemma 6.50.

If $\pi \cap \{i,j\} = \emptyset$ then, by Lemma 6.52, $(a_\pi f)a_{ij} = a_{\pi'}f$, where π' is the ordered set obtained by joining π and $\{i,j\}$. Again by Lemma 6.50 $(a_{\pi'}f)a_{ij} = 0$.

Let $|\pi|$ be odd. If there exists $1 \leq k \leq d-1$ such that $k \in \pi$, $k \notin \{i,j\}$ then $a_\pi = \pm a_{\pi'}v_k$, where $\pi' = \pi - \{k\}$ and therefore $(a_\pi f)R(a_{ij})^2 = \pm(a_{\pi'}f)R(a_{ij})^2R(v_k) = 0$.

If $\pi \cap \{1,2,\ldots,d-1\} \subseteq \{i,j\}$ then it follows that $i,j \in \{1,2,\ldots,d-1\}$ and $\pi = \{i,j,d\}$. In this case $a_\pi = \pm a_{ij}v_d$ and $(a_\pi f)R(a_{ij})^2 = (v_d f)R(a_{ij})^3 = 0$. Lemma is proved. □

Now we can improve slightly Lemma 6.50.

LEMMA 6.54. *Let π, τ be ordered subsets of $\{1,2,\ldots,d\}$ each containing ≥ 2 elements, $|\pi|$ is even and $\pi \cap \tau \neq \emptyset$. Then for arbitrary elements $f, g \in \widehat{Z}$, we have* $(a_\pi f)(a_\tau g) = 0$.

PROOF. If $|\tau|$ is even, then we are in the situation of Lemma 6.50. Let $|\tau|$ be odd. Suppose that there exists an element $i \in \tau$, $1 \leq i \leq d-1$, $i \notin \pi$. Since $v_i^2 = 1$ and $a_\pi Z$ belongs to eigenvalue 1 with respect to $U(v_i)$ it follows that $D(a_\pi Z, v_i) = (0)$. We have $a_\tau g = (a_{\tau'}g)v_i$, where $\tau' = \tau - \{i\}$ and $(a_\pi f)((a_{\tau'}g)v_i) = ((a_\pi f)(a_{\tau'}g))v_i = 0$. If such an i does not exist then $\tau \cap \{1,2,\ldots,d-1\} \subseteq \pi$.

Since $|\tau| \geq 3$ the intersection $\tau \cap \{1,2,\ldots,d-1\}$ contains at least two elements i,j. Then $a_\pi f = (a_{\pi'}f)a_{ij}$ and $(a_\tau g) = (a_{\tau'}g)a_{ij}$ where $\pi' = \pi - \{i,j\}$, $\tau' = \tau - \{i,j\}$. Now $2(a_{\pi'}f)R(a_{ij})R((a_{\tau'}g)a_{ij}) = (a_{\pi'}f)(-R(a_{\tau'}g)R(a_{ij}^2) + R(a_{ij})^2R(a_{\tau'}g) + R(a_{\tau'}g)R(a_{ij})^2 + R(((a_{\tau'}g)a_{ij})a_{ij}) = 0$ by Lemma 6.53 (2). Lemma is proved. □

Denote $\hat{A}_{even} = \widehat{Z} + \sum a_\pi \widehat{Z}$, where the sum runs over all nonempty subsets $\pi \subseteq \{1,2,\ldots,d\}$ such that $|\pi|$ is even.

LEMMA 6.55. $D(\hat{A}_{even}, \hat{A}_{even}) = (0)$.

PROOF. Let $f, g \in \widehat{Z}$, $\emptyset \neq \pi \subseteq \{1,2,\ldots,d\}$, $|\pi|$ is even. Then $D(f, a_\pi g) = 0$ by Lemma 6.52

Let π, τ be two nonempty subsets of $\{1,2,\ldots,d\}$, $|\pi|$ and $|\tau|$ are even. Suppose that there exists $1 \leq i \leq d-1$, $i \in \tau$, $i \notin \pi$. Let $\pi' = \pi \cup \{i\}$. Then

$$D(a_\pi f, a_\tau g) = D((a_{\pi'}f)v_i, a_\tau g) = D(a_{\pi'}f, (a_\tau g)v_i) + D(v_i, (a_{\pi'}f)(a_\tau g)) = 0$$

by Lemmas 6.51 and 6.54. If such an i does not exist then $\tau \cap \{1,2,\ldots,d-1\} \subseteq \pi$ and similarly we can assume that $\pi \cap \{1,2,\ldots,d-1\} \subseteq \tau$, so $\tau \cap \{1,2,\ldots,d-1\} = \pi \cap \{1,2,\ldots,d-1\}$. Since both $|\pi|$ and $|\tau|$ are even it follows that $\pi = \tau$, in which case the assertion is obvious. Lemma is proved. □

If d is odd, then a nonzero element from $\widehat{I}$ can be of weight 1. Indeed, $a_{12\cdots d} \in \widehat{I}(1)$.

Let $w = R(v_1)R(x)R(v_2)R(x)\cdots R(v_d)$. From Lemma 6.48 it follows that $a_{12\cdots d}w = f \bmod \widehat{I}$, $0 \neq f \in \widehat{Z}$.

Suppose that $t^n w = g \bmod \widehat{I}$, $g \in \widehat{Z}$. We have $(t^n - a_{12\cdots d}(f^{-1}g))w \in \widehat{I}$.

The element $(t^n - a_{12\cdots d}(f^{-1}g))w$ is not necessarily homogeneous. Let q be the component of degree n of $t^n - a_{12\cdots d}(f^{-1}g)$. The operator w may seem non-homogeneous because it involves $v_i's$, but in fact it is homogeneous. If $v_i = \frac{1}{\sqrt{2}}(v+u)$, $v_j = \frac{1}{\sqrt{-2}}(v-u)$ are involutions, $v^2 = u^2 = 0$, then $R(v_i)R(x)R(v_j) = \frac{1}{2}(R(v)R(x)R(u) - R(u)R(x)R(v))$. This implies that $qw \in \widehat{I}$. We have $t^n = q \bmod \widehat{I}$.

Clearly, the element q is invertible.

LEMMA 6.56. *Let $h_1, h_2 \in F[q, q^{-1}]$ and $h_1 w, h_2 w \in \widehat{I}$. Then $(h_1 h_2)w \in \widehat{I}$*

PROOF. As in the previous proofs

$$(h_1h_2)D_1 \cdots D_{d-1}R(v_d) = \sum \pm((h_1D_{i_1} \cdots D_{i_s})(h_2D_{j_1} \cdots D_{j_r}))R(v_d),$$

$i_1 < \cdots < i_s$; $j_1 < \cdots < j_r$; $i_1, \ldots, i_s, j_1, \ldots, j_r$ is a permutation of $1, 2, \ldots, d-1$. Let $tr((h_1D_{i_1} \cdots D_{i_s})(h_2D_{j_1} \cdots D_{j_r})R(v_d)) \neq 0$.

Suppose at first that s and r are even Then $h_1D_{i_1} \cdots D_{i_s}$, $h_2D_{j_1} \cdots D_{j_r} \in \hat{A} - \widehat{I}$. These two elements have weights $v_{i_1} \cdots v_{i_s}, v_{j_1} \cdots v_{j_r}$ respectively. This is possible only if one of s, r is equal to 0, and the other one to $d-1$.

Let $r = 0$. Then $tr(h_1D_1 \cdots D_{d-1}R(h_2)R(v_d)) \neq 0$. Since $tr(\hat{A}D(\hat{A}, \hat{A})) = (0)$ it follows that $tr(h_1D_1 \cdots D_{d-1}R(h_2)R(v_d)) = tr(h_1D_1 \cdots D_{d-1}R(v_d)R(h_2)) = tr(h_1wR(h_2)) = 0$ because $h_1w \in \widehat{I}$, the contradiction.

Now let s, r be odd. We will show that both $h_1D_{i_1} \cdots D_{i_s}$ and $h_2D_{j_1} \cdots D_{j_r}$ lie in $\widehat{M_I}$.
If $s = 1$ then $h_1D_{i_1} \in \widehat{M_I}$ by Lemma 6.42. If $s \geq 3$ then $h_1D_{i_1} \cdots D_{i_{s-1}}$ has weight $v_{i_1} \cdots v_{i_{s-1}}$ and therefore lies in $\widehat{I}$. Lemma is proved. □

LEMMA 6.57. *$q^{-1}w \in \widehat{I}$.*

PROOF. Arguing as in the proof of the previous lemma we see that $0 = 1w = (q^{-1}q)w = (q^{-1}w)q \bmod \widehat{I}$. This implies that $q^{-1}w \in \widehat{I}$. Lemma is proved. □

Lemmas 6.56 and 6.57 imply

LEMMA 6.58. *$F[q^{-1}, q]w \subseteq \widehat{I}$.*

If d is even we still denote $q = t^n$.
Let d be odd. There exists an element $h \in \widehat{Z}$ such that $q = t^n - a_{1\cdots d}h$. We shall find an element $h' \in \widehat{Z}$ such that

$$(v_d + v_d(a_{12\cdots d}h'))^2 = q.$$

The equality that we need is

$$2a_{12\cdots d}h'R(v_d)^2 = -a_{12\cdots d}h.$$

Since $a_{12\cdots d} = a_{12\cdots(d-1)}v_d$ we have

$$2a_{12\cdots d}h'R(v_d)^2 = a_{12\cdots(d-1)}h'(-R(v_d^3) + 3R(v_d)R(t^n))$$

Since $D(a_{12\cdots(d-1)}h', t^n) = 0$ by Lemma 6.52, the right hand side is equal to $2a_{12\cdots(d-1)}h'R(v_d)R(t^n)$, so it is sufficient to choose $h' = -\frac{1}{2}t^{-n}h$.

Let $(a_{12\cdots d}h')_0$ be the component of $(a_{12\cdots d}h')$ of degree 0 and let $v_d' = v_d + v_d(a_{12\cdots d}h')_0$. Then ${v_d'}^2 = q$. The subalgebra $F[q^{-1}, q] + \sum_{i=1}^{d-1} v_iF[q^{-1}, q] + v_d'F[q^{-1}, q]$ is isomorphic to $\mathcal{L}(G)$. We could have started with it when splitting A. Hence, without loss of generality we will assume that $t^n w \in \widehat{I}$.

LEMMA 6.59. *For an arbitrary $0 \neq f \in \widehat{Z}$, arbitrary numbers $1 \leq i \neq j \leq d$, $j \leq d-1$, we have $0 \neq (fa_{ij})R(x)R(v_j)R(x) \in \widehat{Z}v_i$.*

PROOF. We know that $(fa_{ij})R(x)R(v_j)R(x) = gv_i + u$, where $0 \neq g \in \hat{Z}$, $u \in \widehat{I}(v_i)$. Our aim is to prove that $u = 0$. If d is even, then $\hat{I}(v_i) = (0)$. Let d be odd.

Denote $w = R(x)R(v_1)R(x)\cdots R(v_{i-1})R(x)R(v_{i+1})R(x)\cdots R(x)R(v_d)$. If $u \neq 0$ then $tr(uw) \neq 0$.

Let us show that $(gv_i)w \in \widehat{I}$. Clearly, $(gv_i)w = (gv_i)D_1D_{i-1}D_{i+1}\cdots D_d$ mod $\widehat{I}$. Since $(gv_i)^2 \in \widehat{Z}$ it follows that $(gv_i)^2D_1\cdots D_{i-1}D_{i+1}\cdots D_d \in \widehat{I}$.

Now, $(gv_i)^2D_1\cdots D_{i-1}D_{i+1}\cdots D_d = \sum \pm(gv_iD_{\mu_1}\cdots D_{\mu_s})(gv_iD_{\nu_1}\cdots D_{\nu_q})$.

If $s \neq 0$, $q \neq 0$ then $(gv_iD_{\mu_1}\cdots D_{\mu_s})(gv_iD_{\nu_1}\cdots D_{\nu_q}) \in \widehat{I}$. Hence, $(gv_i)^2D_1\cdots D_{i-1}D_{i+1}\cdots D_d = 2(gv_i)(gv_iD_1\cdots D_{i-1}D_{i+1}\cdots D_d)$ mod $\widehat{I}$.

Comparing weights we see that if $gv_iD_1\cdots D_{i-1}D_{i+1}\cdots D_d \notin \widehat{I}$ then $gv_iD_1\cdots D_{i-1}D_{i+1}\cdots D_d = h$ mod $\widehat{I}$, $0 \neq h \in \widehat{Z}$. Hence $(gv_i)h \in \widehat{I}$, the contradiction.

Let's show that $(fa_{ij})R(x)^2 \in \widehat{Z}a_{ij}$.

If $d = 3$ then the assertion easily follows from weight considerations. Suppose therefore that $d \geq 5$. Let $k_1 < \cdots < k_{d-2}$ be all numbers from 1 to d, except i and j. Comparing weights we see that $(fa_{ij})R(x)^2 = f'a_{ij} + f''a_{k_1\cdots k_{d-2}}$, where $f', f'' \in \widehat{Z}$. Let $w' = R(v_{k_1})R(x)\cdots R(v_{k_{d-3}})R(x)R(v_{k_{d-2}})$ and $w'' = R(v_{k_1})R(x)\cdots R(v_{k_{d-3}})R(x)$. If $f'' \neq 0$ then $tr((f'a_{ij} + f''a_{k_1\cdots k_{d-2}})w') \neq 0$ and therefore $(f'a_{ij} + f''a_{k_1\cdots k_{d-2}})w'' \notin \widehat{I}$.

We'll check that $(fa_{ij})R(x)^2w'' \in \widehat{I}$.

Indeed, since $k_{d-3} < d$ we have $(fa_{ij})R(x)^2w'' = (fa_{ij})w''R(x)^2 \in \widehat{I}$ by Lemma 6.48. Hence $f'' = 0$, $(fa_{ij})R(x)^2 = f'a_{ij}$.

Now we are ready to show that $(fa_{ij})R(x)R(v_j)R(x)w \in \widehat{I}$. Indeed $(fa_{ij})R(x)R(v_j)R(x)w =$
$(fa_{ij})R(x)R(v_j)R(x)^2R(v_1)R(x)\cdots R(v_{i-1})R(x)R(v_{i+1})R(x)\cdots R(v_d)$.

Furthermore, $(fa_{ij})R(x)R(v_j)R(x)^2R(v_1)R(x) =$
$(f'a_{ij})R(x)R(v_j)R(v_1)R(x) = (f'a_{i1})R(x)^2 = ha_{i1}$, $h \in \widehat{Z}$.

Now,

$$(ha_{i1})R(v_2)R(x)\cdots R(v_{i-1})R(x)R(v_{i+1})\cdots R(x)R(v_d) \notin \widehat{I},$$

which contradicts Lemma 6.48. Lemma is proved. □

Multiplying a_{21} by a proper element from $\widehat{Z}$ we will assume that $a_{21}R(x)R(v_1)R(x) = v_2$. This implies that for any $i \neq j,\ j \leq d-1$, we have $a_{ij}R(x)R(v_j)R(x) = v_i$.

DEFINITION 6.60. We say that an element $y \in \widehat{M}$ satisfies the condition (*) if $[y\widehat{A}, \widehat{M}] \subseteq \widehat{I}$.

Recall that the set $\tilde{W}$ consists of operators

$$R(x)R(v_{i_1})R(x)\cdots R(x)R(v_{i_{2r+1}}),$$

$1 \leq i_1 < \cdots < i_{2r+1} \leq d$ and

$$R(v_{i_1})R(x)R(v_{i_2})\cdots R(x)R(v_{i_{2r}}),$$

$1 \leq i_1 < \cdots < i_{2r} \leq d$.

By Lemma 6.34, for an arbitrary element $y \in M$ satisfying (*) there exists an operator $w \in \tilde{W}$ such that $tr(yw) \neq 0$.

LEMMA 6.61. *Suppose that $d \geq 3$ and $v_i^2 = 1,\ 1 \leq i \leq 3$ (which may not be the case if $d = 3,\ v_3^2 = t^n$). Then $xa_{123} = 0$.*

PROOF. We will show that the element $y = xa_{123}$ satisfies (*) and that for an arbitrary operator $w \in \tilde{W}$ we have $tr(yw) = 0$. Indeed, if $a_{123}R(M)R(A)R(M) \not\subseteq \widehat{I}$ then arguing as in the prood of Lemma 6.32 we conclude that for some $1 \leq k \leq d$ the element $a_{123}R(x)R(v_k)R(x)$ does not lie in $\widehat{I}$ which contradicts the definition of a_{12}. Hence y satisfies (*).

For an operator $w = R(v_{i_1})R(x)R(v_{i_2})\cdots R(x)R(v_{i_{2r}}) \in \tilde{W}$ we have $tr(yw) = tr(a_{123}R(x)w) = 0$ by Lemma 6.48 (2).
Let $w = R(x)R(v_{i_1})R(x)\cdots R(x)R(v_{i_{2r+1}})$.

Since $a_{123}R(x)^2 \in \widehat{Z}a_{123}$ (see the proof of Lemma 6.59) it follows from Lemma 6.48 (2) that $r = 1$ and $\{i_1, i_2, i_3\} = \{1, 2, 3\}$.

Now $a_{123}R(x)^2R(v_1)R(x)R(v_2)R(x)R(v_3) =$
$a_{123}R(v_1)R(x)R(v_2)R(x)R(v_3)R(x)^2 =$
$a_{23}R(x)R(v_2)R(x)R(v_3)R(x)^2 = (-1)R(x)^2 = 0$. Lemma is proved. □

LEMMA 6.62. *Under the assumptions of the previous lemma*
(1) $D(x, a_{12}) = 0$,
(2) if $v_i^2 = v_j^2 = 1$ then $D(x, a_{ij}) = 0$.

PROOF. (1) $a_{12} = a_{123}v_3$. Hence $D(x, a_{12}) = D(x, a_{123}v_3) = D(xa_{123}, v_3) + D(xv_3, a_{123}) = 0$.

(2) Suppose that $\{1, 2\} \cap \{i, j\} = \emptyset$. Then
$D(x, a_{ij}) = [[D(x, a_{12}), D(v_1, v_i)], D(v_2, v_j)] = 0$. If $i = 1,\ j \neq 2$ then $D(x, a_{1j}) = [D(x, a_{12}), D(v_2, v_j)] = 0$. Lemma is proved. □

LEMMA 6.63. *For arbitrary integers $1 \leq i \neq j \leq d$ we have $xR(a_{ij})^2 = 0$.*

PROOF. Suppose that $y = xR(a_{ij})^2 \neq 0$. Clearly, the element y satisfies (*). Comparing weights we see that the only operator $w \in \tilde{W}$ such that $tr(yw) \neq 0$ is $w = R(x)R(v_1)R(x)R(v_2)\cdots R(x)R(v_d)$.

We have $tr(yw) = tr(xR(a_{ij})R(a_{ij})D_1 \cdots D_d)$ and
$xR(a_{ij})R(a_{ij})D_1 \cdots D_d = \sum \pm (xD_{\mu_1} \cdots D_{\mu_s})R(a_{ij}D_{\nu_1} \cdots D_{\nu_q})R(a_{ij}D_{\xi_1} \cdots D_{\xi_p})$, where $\mu_1, \ldots, \mu_s, \nu_1, \ldots, \nu_q, \xi_1, \ldots, \xi_p$ is a permutation of $1, 2, \ldots, d$.

Suppose that $tr((xD_{\mu_1} \cdots D_{\mu_s})R(a_{ij}D_{\nu_1} \cdots D_{\nu_q})R(a_{ij}D_{\xi_1} \cdots D_{\xi_p})) \neq 0$. Then $a_{ij}D_{\xi_1} \cdots D_{\xi_p} \notin \widehat{I} \cup \widehat{M_I}$. This is possible only if $p = 2, \{\xi_1, \xi_2\} = \{i, j\}$. But then $a_{ij}D_{\nu_1} \cdots D_{\nu_q} \in \widehat{I} \cup \widehat{M_I}$. Lemma is proved. □

LEMMA 6.64. *Let $1 \leq i \neq j \leq d-1$. Then $xR(a_{ijd})R(a_{ij}) = 0$.*

PROOF. From the Jordan identity it follows that
$2R(a_{ij}v_d)R(a_{ij}) = -R(a_{ij}^2)R(v_d) + R(a_{ijd}a_{ij}) + R(a_{ij})^2R(v_d) + R(v_d)R(a_{ij})^2$.
Hence we need only to prove $xR(v_d)R(a_{ij})^2 = 0$.

We have $v_d = a_{di}R(x)R(v_i)R(x)$. Hence

$$v_dR(x)R(a_{ij})^2 = a_{di}R(x)R(v_i)R(x)R(x)R(a_{ij})^2 = a'_{di}R(x)R(v_i)R(a_{ij})^2,$$

where $a'_{di} = a_{di}R(x)^2$.

By the Jordan identity $R(v_i)R(a_{ij})^2 = -R(a_{ij})^2R(v_i)$ and for the same reason $R(a'_{di})R(a_{ij})^2 = -R(a_{ij})^2R(a'_{di})$. Hence
$a'_{di}R(x)R(v_i)R(a_{ij})^2 = xR(a_{ij})^2R(a'_{di})R(v_i) = 0$. Lemma is proved. □

LEMMA 6.65. *Let $1 \leq i, j, k, l \leq d$ be distinct integers. The expression $xR(a_{ij})R(a_{kl})$ is skew-symmetric in i, j, k, l.*

PROOF. We have to prove that $x(R(a_{ij})R(a_{kl}) + R(a_{il})R(a_{kj})) = 0$.

Suppose that $j < d$. We have $a_{kl} = a_{kj}D$ and $a_{il} = a_{ij}D$, where $D = D(v_j, v_l)$. Now, $x(R(a_{ij})R(a_{kl}) + R(a_{il})R(a_{kj})) = (xR(a_{ij})R(a_{kj}))D - (xD)R(a_{ij})R(a_{kj})$.

Let us show that $xR(a_{ij})R(a_{jl}) = 0$. If $k < d$, then by Lemma 6.62 (remark that in our case $d \geq 4$) $D(x, a_{kj}) = 0$ and therefore $xR(a_{ij})R(a_{jl}) = xR(a_{ij}a_{kj}) = 0$.

Let $k = d$. Then $a_{dj} = -a_{ijd}v_i$, $xR(a_{ij})R(a_{ijd}v_i) =$
$x(-R(a_{ijd})R(a_{ij}v_i) - R(v_i)R(a_{ij}a_{ijd}) + R(a_{ijd})R(a_{ij})R(v_i) +$
$R(v_i)R(a_{ij})R(a_{ijd}) + R((a_{ijd}v_i)a_{ij})) = 0$ by Lemma 6.64.

Now let us show that $xD(v_j, v_l)R(a_{ij})R(a_{kj}) = 0$ Here we can assume that $l = d$ and therefore $i, j, k < d$. We have $a_{ij} = a_{ijk}v_k$. By the Jordan identity
$R(a_{ijk}v_k)R(a_{kj}) = -R(a_{ijk}a_{kj})R(v_k) - R(v_ka_{kj})R(a_{ijk}) +$
$R(a_{ijk})R(a_{kj})R(v_k) + R(v_k)R(a_{kj})R(a_{ijk}) + R((a_{ijk}v_k)a_{kj}) =$
$R(a_{ijk})R(a_{kj})R(v_k) + R(v_k)R(a_{kj})R(a_{ijk})$.

Now $xD(v_j, v_d)R(v_k) = xR(v_k)D(v_j, v_d) = 0$. By Lemmas 6.61 and 6.62 $xDR(a_{ijk})R(a_{kj}) = xR(a_{ikd})R(a_{kj}) = xR(a_{ikd}a_{kj})$. Now $a_{ikd}a_{kj} = 0$ by Lemma 6.54. Lemma is proved. □

COROLLARY 6.66. For arbitrary distinct integers $1 \leq i_1, \ldots, i_{2r} \leq d$ the expression $x_{i_1 \cdots i_{2r}} = xR(a_{i_1 i_2})R(a_{i_3 i_4}) \cdots R(a_{i_{2r-1} i_{2r}})$ is skew-symmetric in $i_1, i_2, \ldots, i_{2r}$.

LEMMA 6.67. *Let $1 \leq i, j, k \leq d$ be distinct integers such that $j, k \leq d-1$. Then $xR(a_{ij})R(v_j) = xR(a_{ik})R(v_k)$.*

PROOF. $xR(a_{ik})R(v_k) = xR(a_{ij}D(v_j, v_k))R(v_k) = xR(a_{ij})D(v_j, v_k)R(v_k) = -xR(a_{ij})R(v_k D(v_j, v_k)) = xR(a_{ij})R(v_j)$. Lemma is proved. □

Unless $d = 2$ and $v_2^2 = t^n$, for an arbitrary integer $1 \leq i \leq d$ we choose $1 \leq j \leq d-1$ such that $v_j^2 = 1$, $i \neq j$ and denote $[xv_i] = xR(a_{ij})R(v_j)$.

If $d = 2$, $v_2^2 = t^n$ this definition does not work for $[xv_1]$. In this special case we define $[xv_1] = xR(a_{12})R(v_2)$.

Let $1 \leq i_1, \ldots, i_{2r+1} \leq d$ be distinct integers.

Define $x_{i_1, \ldots, i_{2r+1}} = [xv_{i_1}]R(a_{i_2 i_3}) \cdots R(a_{i_{2r} i_{2r+1}})$.

LEMMA 6.68. *$x_{i_1, \ldots, i_{2r+1}}$ is skew-symmetric in $i_1, \ldots, i_{2r+1}$.*

PROOF. It is sufficient to prove that $[xv_i]R(a_{jk})$ is skew-symmetric in i, j, k, that is, $[xv_i]R(a_{jk}) + [xv_j]R(a_{ik}) = 0$.

Suppose at first that $k < d$. Then $[xv_i] = xR(a_{ik})R(v_k)$, $[xv_j] = xR(a_{jk})R(v_k)$ and the equality above becomes $xR(a_{ik})R(v_k)R(a_{jk}) +$
$xR(a_{jk})R(v_k)R(a_{ik}) = 0$ which is an immediate corollary of the Jordan identity.

If $k = d$ then $i < d$ and $j < d$. In this case we have to prove that

$xR(a_{ij})R(v_j)R(a_{jd}) + xR(a_{ji})R(v_i)R(a_{id}) = 0$.

But $xR(a_{ij})R(v_j)R(a_{jd}) = -xR(a_{jd})R(v_j)R(a_{ij})$ and $xR(a_{ji})R(v_i)R(a_{id}) = xR(a_{id})R(v_i)R(a_{ij})$.

Now it remains to notice that $xR(a_{jd})R(v_j) = xR(a_{id})R(v_i)$ by Lemma 6.67. Lemma is proved. □

LEMMA 6.69. *For arbitrary elements $f, g \in \widehat{Z}$ we have $x(R(f)R(g) - R(fg)) \in \widehat{M_I}$.*

PROOF. We have $xR(f)R(g) = xR(f)R((gv_1)v_1) =$
$x(-R(gv_1)R(fv_1) + R(gv_1)R(f)R(v_1) - R(v_1)R(f(gv_1)) + R(v_1)R(f)R(gv_1) + R(fg) = xR(fg) \bmod \hat{M}_I$ by Lemma 6.42 (recall that $\hat{M}_I$ is a subbimodule of $\hat{M}$). Lemma is proved. □

LEMMA 6.70. *For arbitrary elements $y \in \widehat{M_I}$, $f \in \widehat{Z}$ we have*
$yR(x)R(f) = yR(f)R(x) = yR(fx) \bmod \widehat{I}$.

PROOF. Notice that $D(x, f) = D(x, (fv_1)v_1) = D(x(fv_1), v_1) \in D(\widehat{M_I}, v_1)$. Hence $yD(x, f) \in \widehat{M_I}D(\widehat{M_I}, v_1) \subseteq \widehat{I}$ by Lemma 1.141. This proves the first equality.

To prove the second equality we can assume that $y = y'R(a)R(b)$, where $y' \in \widehat{M}$, $a \in \widehat{I}$ and the element b either lies in $\widehat{I}$ or in $\widehat{Z}$ or $b = v_i g$, $g \in \widehat{Z}$. If $b \in \widehat{I}$

or $b \in \widehat{Z}$ then $y'R(a)R(b)R(xf)$ and $y'R(a)R(b)R(x)R(f)$ both lie in $\widehat{I}$ by Lemmas 1.34 and 1.41. Suppose therefore that $b = v_i g$. Then

$y'R(a)R(v_ig)R(xf) = y'R(a)(-R(x)R(v_igf) - R(f)R(x(v_ig)) + R(v_ig)R(x)R(f) + R(f)R(x)R(v_ig) + R(x(v_igf)) = y'R(a)R(v_ig)R(x)R(f) \bmod \widehat{I}$. Lemma is proved. □

LEMMA 6.71. *For arbitrary elements $y \in \widehat{M_I}$, $f \in \widehat{Z}$, arbitrary number $1 \leq i \leq d$, we have $yR(v_if)R(x) = yR(v_i)R(x)R(f)$ mod $\widehat{I}$.*

PROOF. By the Jordan identity $yR(v_if)R(x) = y(-R(xf)R(v_i) - R(xv_i)R(f) + R(f)R(x)R(v_i) + R(v_i)R(x)R(f) + R(x(fv_i))$.

By the previous lemma $yR(xf)R(v_i) = yR(f)R(x)R(v_i) \bmod \widehat{I}$ and by Lemma 6.42 $yR(x(fv_i)) \subseteq [\widehat{M_I}, \widehat{M_I}] \subseteq \widehat{I}$. By Lemma 6.42 $xv_i \in \hat{M}_I$ and therefore $yR(xv_i) \in [\hat{M}_I, \hat{M}_I] \subseteq \hat{I}$. Hence $yR(v_if)R(x) = yR(v_i)R(x)R(f) \bmod \widehat{I}$. Lemma is proved. □

LEMMA 6.72. *For arbitrary distinct integers $1 \leq i_1, \ldots, i_{2r} \leq d$ we have $x_{i_1 \cdots i_{2r}} = xa_{i_1 \cdots i_{2r}}$.*

PROOF. If r = 1 there is nothing to prove. That's why we can assume that $r \geq 2$ and therefore $d \geq 4$. Moreover, since both $x_{i_1 \cdots i_{2r}}$ and $a_{i_1 \cdots i_{2r}}$ are skew-symmetric in $i_1, \ldots, i_{2r}$ we can assume that $i_{2r-1} < d$, $i_{2r} < d$. Now $x_{i_1 \cdots i_{2r}} = x_{i_1 \cdots i_{2r-2}} a_{i_{2r-1} i_{2r}} = (xa_{i_1 \cdots i_{2r-2}}) a_{i_{2r-1} i_{2r}}$ by the induction assumption.

Since $D(x, a_{i_{2r-1} i_{2r}}) = 0$ by Lemma 6.62, we have $(xa_{i_1 \cdots i_{2r-2}}) a_{i_{2r-1} i_{2r}} = x(a_{i_1 \cdots i_{2r-2}} a_{i_{2r-1} i_{2r}}) = xa_{i_1 \cdots i_{2r}}$. Lemma is proved. □

Let $\tilde{W}_1$ be the set of operators of the types

$$R(v_{i_1})R(x)R(v_{i_2})R(x) \cdots R(x)R(v_{i_{2r}}),$$

where $1 \leq i_1, \ldots, i_{2r} \leq d$ are distinct integers and

$$R(x)R(v_{i_1})R(x)R(v_{i_2})R(x) \cdots R(x)R(v_{i_{2r+1}}),$$

where $1 \leq i_1, \ldots, i_{2r+1} \leq d$ are distinct integers.

LEMMA 6.73. *For an operator $w \in \tilde{W}_1$ we have $tr(x_{i_1 \cdots i_{2r}} w) \neq 0$ if and only if $w = R(v_{j_1})R(x)R(v_{j_2})R(x) \cdots R(x)R(v_{j_{2r}})$, where $j_1, \ldots, j_{2r}$ is a permutation of $i_1, \ldots, i_{2r}$.*

PROOF. Immediatly follows from Lemmas 6.72 and 6.48 (1). □

LEMMA 6.74. *For arbitrary distinct integers $1 \leq i, j, k \leq d$, we have $D([xv_i], a_{jk}) = 0$.*

PROOF. At least one of the integers j, k is distinct from d. Let $j < d$. Then $[xv_i] = xR(a_{ij})R(v_j)$ and $D([xv_i], a_{jk}) = D((xa_{ij})v_j, a_{jk}) = D(v_j, xR(a_{ij})R(a_{jk})) = 0$, because $xR(a_{ij})R(a_{jk}) = 0$ as shown in the proof of Lemma 6.65. Lemma is proved. □

COROLLARY 6.75. $x_{i_1 \cdots i_{2r+1}} = [xv_{i_1}] a_{i_2 \cdots i_{2r+1}}$.

Remark Let $1 \le \mu, i_1, \dots, i_{2r+1} \le d$ be distinct integers and $v_\mu^2 = 1$. Since $a_{i_2 \cdots i_{2r+1}}$ belongs to eigenvalue 1 with respect to $U(v_\mu)$ it follows that $D(a_{i_2 \cdots i_{2r+1}}, v_\mu) = 0$. Then $x_{i_1 \cdots i_{2r+1}} = [xv_{i_1}]a_{i_2 \cdots i_{2r+1}} = xR(a_{i_1 \mu})R(v_\mu)R(a_{i_2 \cdots i_{2r+1}}) = x_{i_1 \cdots i_{2r+1}\mu} v_\mu$.

LEMMA 6.76. *For an operator $w \in \tilde{W}_1$ we have $tr(x_{i_1 \cdots i_{2r+1}} w) \neq 0$ if and only if $w = R(x)R(v_{j_1})R(x) \cdots R(x)R(v_{j_{2r}})R(x)R(v_{j_{2r+1}})$ where $j_1, \dots, j_{2r+1}$ is a permutation of $i_1, \dots, i_{2r+1}$.*

PROOF. Suppose at first that there exists μ such that $\mu \notin \{i_1, \dots, i_{2r+1}\}$ and $v_\mu^2 = 1$. Then $x_{i_1 \cdots i_{2r+1}} = xR(a_{i_1 \cdots i_{2r+1}\mu})R(v_\mu)$.

If $w = R(x)R(v_{j_1})R(x) \cdots R(x)R(v_{j_{2s+1}})$, then $x_{i_1 \cdots i_{2r+1}} w = a_{i_1 \cdots i_{2r+1}\mu} R(x)R(v_\mu)R(x)R(v_{j_1}) \cdots R(v_{j_{2s+1}})$.

If $\mu \notin \{j_1, \dots, j_{2s+1}\}$, then $R(x)R(v_\mu)R(x)R(v_{j_1}) \cdots R(v_{j_{2s+1}}) \in W_1$ and we can refer to Lemma 6.48.

If $r = s$ and $\{i_1, \dots, i_{2r+1}\} = \{j_1, \dots, j_{2s+1}\}$ then $\mu \notin \{j_1, \dots, j_{2s+1}\}$. Hence $R(x)R(v_\mu)R(x)R(v_{j_1})R(x) \cdots R(x)R(v_{j_{2s+1}}) \in W_1$ and therefore $t(x_{i_1 \cdots i_{2r+1}} w) \neq 0$ by Lemma 6.48.

If $\mu \notin \{j_1, \dots, j_{2s+1}\}$ but $\{i_1, \dots, i_{2r+1}\} \neq \{j_1, \dots, j_{2s+1}\}$ then $t(x_{i_1 \cdots i_{2r+1}} w) = 0$ by Lemma 6.48.

If $\mu \in \{j_1, \dots, j_{2s+1}\}$ then $R(v_\mu)w = \sum R(v_\mu)R(x)R(v_\mu) \cdots = 0$. Indeed, for an arbitrary $j \neq \mu$, $1 \le j \le d$, we have $R(v_j)R(x)R(v_\mu) = -R(v_\mu)R(x)R(v_j) + R(v_\mu)R(xv_j)$ by the Jordan identity.

Now let $w = R(v_{j_1})R(x)R(v_{j_2}) \cdots R(x)R(v_{j_{2s}})$. We will show that $t(x_{i_1 \cdots i_{2r+1}} w) = 0$.

Suppose that $t(x_{i_1 \cdots i_{2r+1}} w) \neq 0$. Since the element $x_{i_1 \cdots i_{2r+1}} w$ has weight $v_{i_1} \cdots v_{i_{2r+1}} v_{j_1} \cdots v_{j_{2s}}$ it follows that $d = 2r+1+2s$ and $\{j_1, \dots, j_{2s}\} = \{1, 2, \dots, d\} - \{i_1, \dots, i_{2r+1}\}$.

If $\mu = j_1$ then $a_{i_1 \cdots i_{2r+1}\mu} R(x)R(v_\mu)R(v_{j_1})R(x) = a_{i_1 \cdots i_{2r+1}\mu} R(x)^2 \in a_{i_1 \cdots i_{2r+1}\mu} \widehat{Z}$ (see the proof of Lemma 6.59 and use the fact that $a_{i_1 \cdots i_{2k}}$ is a product of $a'_{ij}s$) in which case our assertion follows from Lemma 6.48.

Suppose that $j_1 \neq \mu$. If $j_1 \neq d$ and $a_{i_1 \cdots i_{2r+1}\mu} R(x)R(v_\mu)R(v_{j_1}) \neq 0$ then $a_{i_1 \cdots i_{2r+1}\mu} R(x)R(v_{j_1})R(v_\mu) = 0$. Hence, $a_{i_1 \cdots i_{2r+1}\mu} R(x)R(v_\mu)R(v_{j_1}) = a_{i_1 \cdots i_{2r+1}\mu} R(x)D(v_\mu, v_{j_1}) = a_{i_1 \cdots i_{2r+1}\mu} D(v_\mu, v_{j_1})R(x) = a_{i_1 \cdots i_{2r+1} j_1} R(x)$ and again we can apply Lemma 6.48.

Now suppose that $j_1 = d$ and $a_{i_1 \cdots i_{2r+1}\mu} R(x)R(v_\nu)R(v_d) \neq 0$.

Then $a_{i_1 \cdots i_{2r+1}\mu} R(x)R(v_\nu)R(v_d) = a_{i_1 \cdots i_{2r+1}\mu} R(x)D(v_\mu, v_d) = a_{i_1 \cdots i_{2r+1} d} R(x) + a_{i_1 \cdots i_{2r+1}\mu} R(xD(v_\mu, v_d))$.

We have $t(a_{i_1 \cdots i_{2r+1}\mu} R(x)R(x)R(v_{j_2}) \cdots R(v_{j_{2s}})) = 0$ by Lemma 6.48.

It remains to show that $t(a_{i_1 \cdots i_{2r+1}\mu} R(xD(v_\mu, v_d))D_{j_2} \cdots D_{j_{2s}}) = 0$.

We have $a_{i_1 \cdots i_{2r+1}\mu} R(xD(v_\mu, v_d))D_{j_2} \cdots D_{j_{2s}} = \pm \sum (a_{i_1 \cdots i_{2r+1}\mu} D_{\nu_1} \cdots D_{\nu_r}) R(xD(v_\mu, v_d)D_{p_1} \cdots D_{p_t})$.

By Lemma 6.48 $a_{i_1 \cdots i_{2r+1}\mu} D_{\nu_1} \cdots D_{\nu_r} \in \widehat{I} \cup \widehat{M_I}$ in every summand, a contradiction.

Now suppose that $v_\mu^2 = 1$ implies $\mu \in \{i_1, \dots, i_{2r+1}\}$.

Thus d is odd and $\{i_1, \dots, i_{2r+1}\} = \{1, 2, \dots, d\}$. Comparing weights we see that $tr(x_{1\cdots d}w) \neq 0$, $w \in \tilde{W}_1$ is possible only if
$w = R(x)R(v_{j_1})R(x) \cdots R(x)R(v_{j_d})$, $\{j_1, \dots, j_d\} = \{1, 2, \dots, d\}$ Lemma is proved. □

LEMMA 6.77. *For arbitrary elements* $y \in \widehat{M_I}$, $f \in \widehat{Z}$ *and an arbitrary operator* $w \in \tilde{W}_1$, *we have* $tr((yf)w) = tr(yw)f$.

PROOF. Suppose first that $w = R(x)R(v_{i_1})R(x) \cdots R(x)R(v_{i_{2r+1}})$. Then $tr((yf)w) = tr((yf)D_{i_1} \cdots D_{i_{2r+1}})$; $(yf)D_{i_1} \cdots D_{i_{2r+1}} =$ $\sum \pm (yD_{\mu_1} \cdots D_{\mu_s})(fD_{\nu_1} \cdots D_{\nu_q})$; $\mu_1, \dots, \mu_s, \nu_1, \dots, \nu_q$ is a permutation of $i_1, \dots,$ i_{2r+1}.

Suppose that $tr((yD_{\mu_1} \cdots D_{\mu_s})(fD_{\nu_1} \cdots D_{\nu_q})) \neq 0$. Let q be odd. The element $fD_{\nu_1} \cdots D_{\nu_{q-1}}$ has weight $v_{\nu_1} \cdots v_{\nu_{q-1}}$. This element can lie outside of $\widehat{I}$ only if $q = 1$ or if d is odd and $q = d$. If $q = 1$ then $fD_{\nu_1} \in \widehat{M_I}$ by Lemma 6.42.

If $q = d$ then $s = 0$ in which case $tr[y, fD_1 \cdots D_d] = 0$ by Lemma 1.28 Hence q is even. The element $fD_{\nu_1} \cdots D_{\nu_q}$ has weight $v_{\nu_1} \cdots v_{\nu_q}$. It can lie outside of $\widehat{I}$ only if $q = 0$ or if d is odd and $q = d - 1$. The latter is though impossible because of the assumption made after we proved by Lemma 6.58. Hence $q = 0$.

Now let $w = R(v_{i_1})R(x)R(v_{i_2})R(x) \cdots R(x)R(v_{i_{2r}})$. As above $tr((yf)w) = tr((yf)D_{i_1} \cdots D_{i_{2r-1}}R(v_{i_{2r}}))$ and $(yf)D_{i_1} \cdots D_{i_{2r-1}} = \sum \pm (yD_{\mu_1} \cdots D_{\mu_s})(fD_{\nu_1} \cdots D_{\nu_q})$.

Arguing as above we see that $tr(((yD_{\mu_1} \cdots D_{\mu_s})(fD_{\nu_1} \cdots D_{\nu_q}))v_{i_{2r}}) \neq 0$ is possible only if $q = 0$. Lemma is proved. □

LEMMA 6.78. $\widehat{M} = \sum x_\pi \widehat{Z}$, *where* π *runs over all subsets of* $\{1, 2, \dots, d\}$ *and* $x_\emptyset = x$, $x_i = [xv_i]$.

PROOF. By Lemmas 6.42, 6.69 we have that $\widehat{M} = x\widehat{Z} + \widehat{M_I}$. Suppose that a root element $y \in \widehat{M_I}$ does not satisfy the (*) condition. Moreover, suppose also that there exists an element $y' \in \widehat{M}$ such that $[y, y'] \notin \widehat{I}$. Since $[y, \widehat{M_I}] \subseteq \widehat{I}$ by Lemma 1.41 we can assume that $y' \in x\widehat{Z}$. By Lemma 6.70 $[y, x] \notin \widehat{I}$. Hence, there exists $1 \leq j \leq d$ such that $tr(yR(x)R(v_j)) \neq 0$.

Now suppose that $[y, \widehat{M}] \subseteq \widehat{I}$ but there exist root elements such that $yR(a)R(y') \notin \widehat{I}$. As above, without loss of generality we can assume that $y' = x$. Since $yR(\widehat{I})R(x) \subseteq \widehat{I}$ by Lemma 1.27 we can assume that $a \in \widehat{\mathcal{L}(G)}$.

If $a \in \widehat{Z}$ then by Lemma 6.70 $yR(ax) \notin \widehat{I}$, the contradiction. Hence, $a = v_i f$, $f \in \widehat{Z}$. By Lemma 6.71 $yR(v_i)R(x) \notin \widehat{I}$. Therefore there exists j, $1 \leq j \leq d$ such that $tr(yR(v_i)R(x)R(v_j)) \neq 0$. We can assume $i < j$.

Consider the set of operators

$$\tilde{W}_{ext} = \tilde{W} \cup \{R(x)R(v_j),\ 1 \leq j \leq d\} \cup \{R(v_i)R(x)R(v_j),\ 1 \leq i < j \leq d\}.$$

We showed that for an arbitrary nonzero element $y \in \widehat{M_I}$ there exists an operator $w \in \tilde{W}_{ext}$ such that $tr(yw) \neq 0$. Moreover by Lemmas 6.73 and 6.76 for every operator $w \in \tilde{W}_{ext}$ there exists a nonempty subset $\pi = \pi(w)$ of $\{1, 2, \dots, d\}$ such that $tr(x_{\pi(w)}w) \neq 0$ and $tr(x_\tau w) = 0$ for every $\tau \neq \pi(w)$.

Now we are ready to prove the lemma.

Let y be an arbitrary nonzero element from $\widehat{M_I}$. Let $tr(yw) = g_w$, $w \in \tilde{W}_{ext}$ and let $tr(x_{\pi(w)}w) = f_w \neq 0$. Then by Lemma 6.77
$tr((y - \sum_{w \in \tilde{W}_{ext}} x_{\pi(w)}(f_w^{-1}g_w))w') = 0$ for every $w' \in \tilde{W}_{ext}$.

Hence $y = \sum_{w \in \tilde{W}_{ext}} x_{\pi(w)}(f_w^{-1}g_w)$. Lemma is proved. □

LEMMA 6.79. $[[xv_i], [xv_j]] = a_{ij}$ *unless* $d = 2$, $v_2^2 = t^n$. *In the latter case* $[[xv_i], [xv_j]] = a_{ij}t^n$.

PROOF. Suppose that $i < d$. Then $[xv_i]R([xv_j]) = [xv_i]R((xa_{ji})v_i) =$
$[xv_i](-R(x)R(v_i)R(a_{ji}) - R(a_{ji})R(v_i)R(x) + R(v_i)R(xa_{ji})) =$
$[xv_i]R(x)R(v_i)R(a_{ij})$ because $[xv_i]R(a_{ji}) = 0$ (see the proof of Lemma 6.68) and $[xv_i]R(v_i) = 0$.

Unless $d = 2$, $v_2^2 = t^n$ we have $[xv_i]R(x) = v_i$, which implies the result. If $d = 2$, $v_2^2 = t^n$ then $[xv_1]R(x) = t^n v_1$. Lemma is proved. □

LEMMA 6.80. *For arbitrary integers* $1 \leq i \neq j \leq d$, $j < d$, *arbitrary element* $f \in \widehat{Z}$ *we have* $xR(a_{ij})R(v_j)R(x)R(f) = xR(a_{ij})R(v_j)R(xf) =$
$xR(a_{ij})R(v_jf)R(x) = xR(a_{ij}f)R(v_j)R(x) = xR(f)R(a_{ij})R(v_j)R(x) =$
$xR(a_{ij})R(v_j)R(f)R(x)$.

PROOF. By the Jordan identity $xR(a_{ij})R(v_j)R(x)R(f) =$
$xR(a_{ij})(-R(f)R(x)R(v_j) - R(x(fv_j)) + R(x)R(fv_j) + R(v_j)R(xf))$.

Clearly, $xR(a_{ij})R(f)R(x)$ belongs to eigenvalue -1 with respect to $U(v_j)$, hence $xR(a_{ij})R(f)R(x)R(v_j) = 0$.

Also, $xR(a_{ij})R(x)R(fv_j) \in (a_{ij}\tilde{Z})(fv_j) = 0$ by Lemma 6.51 and the proof of Lemma 6.59.

Let us show that $xR(a_{ij})R(x(fv_j)) = 0$. This element clearly lies in $I(v_i)$. If d is even then $I(v_i) = (0)$. Let d be odd. We have to check that
$tr(xR(a_{ij})R(x(fv_j))R(x)R(v_1)R(x) \cdots R(v_{i-1})R(x)R(v_{i+1}) \cdots R(x)R(v_d)) = 0$, or equivalently , that $tr(xR(a_{ij})R(x(fv_j))D_1 \cdots D_{i-1}D_{i+1} \cdots D_d) =$
$tr(fD_jR(xa_{ij})D_1 \cdots D_{i-1}D_{i+1} \cdots D_d) = 0$.

Let $fD_jR(xa_{ij})D_1 \cdots D_{i-1}D_{i+1} \cdots D_d =$
$\sum \pm(fD_jD_{\mu_1} \cdots D_{\mu_s})R(a_{ij}R(x)D_{\nu_1} \cdots D_{\nu_q})$, where $\mu_1, \dots, \mu_s, \nu_1, \dots, \nu_q$ is a permutation of $1, 2, \dots, i-1, i+1, \dots, d$; $\mu_1 < \cdots < \mu_s$; $\nu_1 < \cdots < \nu_q$.

The element $fD_jD_{\mu_1} \cdots D_{\mu_s}$ can lie outside of $\widehat{I} \cup \widehat{M_I}$ only if $s = 1$, $\mu_1 = j$, in which case $a_{ij}R(x)D_{\nu_1} \cdots D_{\nu_q} \in \widehat{I}$. Other equalities are proved similarly. Lemma is proved. □

If $d = 2$, $v_2^2 = t^n$ then we can assume also $j = 2$ in Lemma 6.80. Indeed, $xR(a_{12})R(v_2)R(x)R(f) - xR(a_{12})R(v_2)R(xf) \in I(v_1) = (0)$ because $d = 2$ is even.

LEMMA 6.81. *If* $d = 2$ *then* $\widehat{M_I}\widehat{I} = (0)$.

PROOF. By Lemma 6.49 we have $\widehat{I} = a_{12}\widehat{Z}$. We will prove the lemma in several steps.

1) By Lemma 6.52 $D(\widehat{Z}, \widehat{I}) = 0$.

2) $JR(a_{12})^2 = (0)$ by Lemma 6.53. Let us show that $\hat{M}R(a_{12})^2 = (0)$. By Lemma 6.78 $\hat{M} = x\hat{Z} + [xv_1]\hat{Z} + [xv_2]\hat{Z}$. From 1) and Lemma 6.63 it follows tht $(x\hat{Z})R(a_{12})^2 = (0)$. It remains to check that $[xv_i]R(a_{12})^2 = 0$. But $[xv_i] = \pm xR(a_{12})R(v_j)$ and $R(a_{12})R(v_j)R(a_{12}) = 0$ because of the Jordan identity and the fact that $a_1v_j = a_{12}^2 = 0$.

3) For an arbitrary $f \in \widehat{Z}$ we have $R(a_{12}f)R(a_{12}) = 0$. Indeed, by the Jordan identity $2R(a_{12}f)R(a_{12}) = -R(a_{12}^2)R(f) + R(a_{12})^2R(f) +$ $R(f)R(a_{12})^2 + R((a_{12}f)a_{12}) = 0$.

4) $D(\widehat{A}, \widehat{I}) - D(\widehat{A}, a_{12})$. Indeed, for arbitrary $f, g \in \widehat{Z}$ we have $(v_if, a_{12}g) = D((v_if)a_{12}, g) + D((v_if)g, a_{12}) = D(v_ifg, a_{12})$.

5) For arbitrary $f, g \in \widehat{Z}$ we have $R(fa_{12})R(ga_{12}) = 0$. Indeed, by the Jordan identity $R(fa_{12})R(ga_{12}) = -R(g)R((fa_{12})a_{12}) - R(a_{12})R((fa_{12})g) +$ $R(a_{12})R(fa_{12})R(g) + R(g)R(fa_{12})R(a_{12}) + R((ga_{12})(fa_{12})) = 0$ by 2).

Hence $R(\widehat{I})R(\widehat{I}) = (0)$.

6) $R(\widehat{I})R < \widehat{A} > R(\widehat{I}) = (0)$. Indeed, by the Jordan identity it is sufficient to prove that $R(\widehat{I})R(\widehat{A})R(\widehat{I}) = (0)$. By 5) and 4) $R(\widehat{I})R(\widehat{A})R(\widehat{I}) = R(\widehat{I})D(\widehat{A}, \widehat{I}) =$ $R(\widehat{I})D(\widehat{A}, a_{12}) = R(\widehat{I})R(\widehat{A})R(a_{12})$.

Again $R(\widehat{I})R(\widehat{A})R(a_{12}) = D(\widehat{I}, \widehat{A})R(a_{12}) = D(a_{12}, \widehat{A})R(a_{12}) =$ $R(a_{12})R(\widehat{A})R(a_{12}) = (0)$ by the Jordan identity and Lemma 6.53(1). Lemma is proved. □

LEMMA 6.82. *For any $1 \leq i \leq d$, $f \in \widehat{Z}$ we have $[xv_i](v_if) = 0$.*

PROOF. Suppose that there exists $k < d, k \neq i$. Then $[x, v_i] = xR(a_{ik})R(v_k)$, and $[xv_i](v_if) = xR(a_{ik})R(v_k)R(v_if) = -xR(v_if)R(v_k)R(a_{ik})$ by the Jordan identity and Lemma 6.51.

Now, $v_if = xR(a_{ik}f)R(v_k)R(x)$. Hence $(v_if)R(x)R(v_k)R(a_{ik}) =$ $xR(a_{ik}f)R(v_k)R(x)^2R(v_k)R(a_{ik}) \in xR(a_{ik}\widehat{Z})R(a_{ik}) = (0)$, as follows from the Jordan identity and the facts that $xR(a_{ik})^2 = 0$ and $(a_{ik}\hat{Z})a_{ik} = (0)$.

Now let $d = 2, i = 1$. Then we have to check that $xR(a_{12})R(v_2)R(v_1f) = 0$. As above, $xR(a_{12})R(v_2)R(v_1f) = -xR(v_1f)R(v_2)R(a_{12}) = 0$ because $M_II = (0)$ by Lemma 6.81. Lemma is proved.

LEMMA 6.83. *For arbitrary $1 \leq i \leq d$, arbitrary elements $f, g \in \widehat{Z}$ we have $D([xv_i]f, g) = 0$.*

PROOF. There exist elements $h', h'' \in \widehat{Z}$ such that $g = (v_ih')(v_ih'')$. Now $D([xv_i]f, (v_ih')(v_ih'')) = D(([xv_i]f)(v_ih'), v_ih'') + D(([xv_i]f)(v_ih''), v_ih') = 0$ by Lemma 6.82. Lemma is proved. □

LEMMA 6.84. *For arbitrary integers $1 \leq i, j, k \leq d$, $j \neq k$ and an arbitrary element $f \in \widehat{Z}$ we have $D([xv_i]f, a_{jk}) = 0$.*

PROOF. Since $j \neq k$, one of these integers is $< d$. Let $j < d$ and $j \neq i$. Then $D(xfR(a_{ij})R(v_j), a_{jk}) = D(xfR(a_{ij}), a_{jk}v_j) + D(v_j, xfR(a_{ij})R(a_{jk})) = 0$ by Lemmas 6.52 and 6.65.

Let $i = j < d$. We have to show that $D([xv_i]f, a_{ik}) = 0$. Suppose that $k < d$. Then $D(xfR(a_{ik})R(v_k), a_{ik}) = 0$ for the same reason as above.

Let $k = d$. We have to show that $D([xv_i]f, a_{id}) = 0$. Suppose that $d \geq 3$. Then we can choose $\mu < d$, $\mu \neq i$. We have $D(xfR(a_{i\mu})R(v_\mu), a_{id}) = D(xfR(a_{i\mu}), a_{\mu id}) = -D((xf)a_{i\mu}, v_d a_{i\mu}) = -D((xf)R(a_{i\mu})R(v_d), a_{i\mu})$.

Let us show that $(xf)R(a_{i\mu})R(v_d) \in x_{i\mu d}\widehat{Z}$. If d is even then the assertion follows from Lemma 6.78. Let d odd and let $\{j_1, \ldots, j_{2s}\} = \{1, 2, \ldots, d\} \setminus \{i, \mu, d\}$. By Lemma 6.78 $(xf)R(a_{i\mu})R(v_d) = x_{i\mu d}h + x_{j_1 \cdots j_{2s}}g$, $h, g \in \hat{Z}$. Suppose that $g \neq 0$. By Lemmas 6.73, 6.77 we have $(x_{i\mu d}h)D_{j_1} \cdots D_{j_{2s-1}} \in I$ and $(x_{j_1 \cdots j_{2s}}g)D_{j_1} \cdots D_{j_{2s-1}} \in A \setminus I$.

Hence $(xf)R(a_{i\mu})R(v_d)D_{i_1} \cdots D_{j_{2s-1}} \in A \setminus I$. By Lemma 6.52 $(xf)R(a_{i\mu})R(v_d) = xR(a_{i\mu})R(v_d)R(f)$. Now by Lemma 6.77 $xR(a_{i\mu})R(v_d)D_{i_1} \cdots D_{j_{2s-1}} \in A \setminus I$.

We have

$$xR(a_{i\mu})R(v_d)D_{i_1} \cdots D_{j_{2s-1}} = \sum \pm((xa_{i\mu})D_{\nu_1} \cdots D_{\nu_q})(v_d D_{\xi_1} \cdots D_{\xi_p}),$$

where $p + q = 2s - 1$ and $\nu_1, \ldots, \nu_q, \xi_1, \ldots, \xi_p$ is a permutation of $j_1, \ldots, j_{2s-1}$.

Suppose that $q \geq 1$. Then $(xa_{i\mu})D_{\nu_1} = a_{i\mu}R(x)^2R(v_{\nu_1}) - [(xa_{i\mu})v_{\nu_1}), x]$. But $a_{i\mu}R(x)^2 = 0$ buy Lemma 6.62(2) and $(xa_{i\mu})v_{\nu_1} = 0$ because $xa_{i\mu}$ belongs to eigenvalue -1 with respect to $U(v_{\nu_1})$.

If $q = 0$, then $[xa_{i\mu}, v_d D_{j_1} \cdots D_{j_{2s-1}}] \in [MI, M] \subseteq I$ by Proposition 1.4(1). This contradiction proves our claim.

Hence $(xf)R(a_{i\mu})R(v_d) = [xv_d]R(a_{i\mu})R(h) = ([xv_d]h)a_{i\mu}$, $h \in \widehat{Z}$.

Now $D(([xv_d]h)a_{i\mu}, a_{i\mu}) = \frac{1}{2}D([xv_d]h, a_{i\mu}^2) = 0$.

Let $d = 2$. Then $D(xfR(a_{12})R(v_1), a_{12}) = D(v_1, xfR(a_{12})^2) = 0$. Lemma is proved. □

LEMMA 6.85. *For arbitrary numbers $1 \leq i, j \leq d$,, arbitrary $f \in \widehat{Z}$ we have $D = D([xv_i]f, [xv_j]g) = 0$.*

PROOF. From Lemmas 6.83, 6.84 it follows that $\widehat{Z}D = 0$ and for arbitrary numbers $1 \leq k \neq l \leq d$, $a_{kl}D = 0$. Now our aim will be to show that $v_k D = 0$ for any k, $1 \leq k \leq d$.

Suppose at first that $k \neq i$, $k \neq j$. The element $v_k D$ has weight $v_k v_i v_j$ and besides $(v_k D)v_k = \frac{1}{2}v_k^2 D = 0$.

This implies (even in the case $k = d$) that $v_k D = 0$.

Suppose now that $k = i$. We have to prove that $v_i D = 0$. By Lemma 6.82, $v_i D = [([xv_j]g)v_i, [xv_i]f]$.

If $j = i$ then again we have only to refer to Lemma 6.82. Suppose therefore that $i \neq j$.

Let $i < d$. Then $[xv_j]R(g)R(v_i) = xR(a_{ij})R(v_i)R(g)R(v_i) = (xa_{ji})g$. By Lemmas 6.83, 6.84, $[(xR(a_{ji})R(g), [xv_i]f] = [x, [xv_i]f]R(a_{ji})R(g)$. By Lemma 6.59 $[x, [xv_i]f] \in \widehat{Z}v_i$ and finally $(\widehat{Z}v_i)a_{ji} = 0$.

The only case that remains open is $v_dD([xv_d]f, [xv_j]g)$, $j \neq d$.

We have $(xv_j)D = (xD)v_j = 0$. But the element xD has eigenvalue 1 with respect to $U(v_j)$. Hence, $xD = 0$.

Now, since $0 \neq xR(a_{dj})R(v_j)R(x) \in \widehat{Z}v_d$ by Lemma 6.59, we conclude that $v_dD = 0$. Since the elements $f \in \hat{Z}$, v_k, a_{ij}, x generate J it follows that $JD = (0)$. Lemma is proved. □

Let S be the subalgebra of $\widehat{J}$ generated by $\widehat{Z}$, $[xv_i]$, $1 \leq i \leq d$.

LEMMA 6.86. $D(S, S) = 0$.

PROOF. Let $X = \{\widehat{Z}, [xv_i], 1 \leq i \leq d\}$. We have to check that $D(X, X) = D(XX, X) = 0$.

By Lemmas 6.83, 6.85 $D(Z, [xv_i]) = (0)$ and $D([xv_i], [xv_j]) = 0$ for arbitrary numbers $1 \leq i, j \leq d$. Hence $D(X, X) = (0)$. To prove that $D(XX, X) = (0)$ we need only to verify that $D([[xv_i], [xv_j]], \hat{Z}) = (0)$, $D([[xv_i], [xv_j]], [xv_k]) = 0$. By Lemma 6.79 $[[xv_i], [xv_j]] = a_{ij}$ or $a_{ij}t^n$. By Lemmas 6.84, 6.52 this implies the result unless $d = 2$.

Let $d = 2$. We have to show that $D(a_{12}t^n, [xv_k]) = 0$, $k = 1$ or 2. By Lemma 6.84 $D(a_{12}t^n, [xv_k]) = D(t^n, [xv_k]a_{12})$, but $[xv_k]a_{12} \in \hat{M}_I I = (0)$ by Lemma 6.81. Lemma is proved. □

LEMMA 6.87. *Let π, $\tau \subseteq \{1, 2, \ldots, d\}$, where $|\pi|$, $|\tau|$ are even, let $f, g \in \widehat{Z}$. Then*

(1)

$$(a_\pi f)(a_\tau g) = \begin{cases} 0 & \text{if } \pi \cap \tau \neq \emptyset \\ a_{\pi\tau} fg & \text{if } \pi \cap \tau = \emptyset. \end{cases}$$

(2) If $1 \leq i_1, \ldots, i_{2r+1} \leq d$ are distinct integers then

$$(x_{i_1 \cdots i_{2r+1}} f)(a_\pi g) = \begin{cases} 0 & \text{if } \{i_1, \ldots, i_{2r+1}\} \cap \pi \neq \emptyset \\ x_{i_1 \cdots i_{2r+1}\pi} fg & \text{otherwise.} \end{cases}$$

(3) If $1 \leq i_1, \ldots, i_{2r+1} \leq d$ are distinct integers and $1 \leq j_1, \ldots, j_{2s+1} \leq d$ are distinct integers, $d \neq 2$ then

$$[x_{i_1 \cdots i_{2r+1}} f, x_{j_1 \cdots j_{2s+1}} g] = \begin{cases} 0 \; if \{i_1, \ldots, i_{2r+1}\} \cap \{j_1 \cdots j_{2s+1}\} \neq \emptyset \\ a_{i_1 \cdots i_{2r+1} j_1 \cdots j_{2s+1}} fg \; otherwise. \end{cases}$$

(4) If $d = 2$, $v_2^2 = t^n$, then $[[xv_1]f, [xv_2]g] = a_{12} fgt^n$.

PROOF. (1) Immediatly follows from Lemmas 6.50 and 6.52.

(2) Suppose that $\{i_1, \ldots, i_{2r+1}\} \cap \pi \neq \emptyset$. In view of Lemmas 6.68, 6.55, 6.74 in order to prove that $(x_{i_1 \cdots i_{2r+1}} f)(a_\pi g) = 0$ we need only to prove that $[xv_i]a_{ij} = 0$,

which was done in the proof of Lemma 6.68. Let $\{i_1,\dots,i_{2r+1}\}\cap\pi=\emptyset$. We have $x_{i_1\cdots i_{2r+1}}=x_{i_1\cdots i_{2r-1}}a_{i_{2r}i_{2r+1}}$.

By Lemma 6.55 $((x_{i_1\cdots i_{2r-1}}f)a_{i_{2r}i_{2r+1}})(a_\pi g)=((x_{i_1\cdots i_{2r-1}}f)(a_\pi g))a_{i_{2r}i_{2r+1}}$.

By the induction assumption about r we have $((x_{i_1\cdots i_{2r-1}}f)(a_\pi g))a_{i_{2r}i_{2r+1}}=(x_{i_1\cdots i_{2r-1}\pi}(fg))a_{i_{2r}i_{2r+1}}=x_{i_1\cdots i_{2r+1}\pi}(fg)$.

(3) and (4) follow from Lemmas 6.55, 6.74, 6.79 and 6.86. Lemma is proved. □

COROLLARY 6.88. $S=\widehat{Z}+\sum a_\pi\widehat{Z}+\sum x_\tau\widehat{Z}$, where π and τ run over all subsets of $\{1,2,\dots,d\}$ of even and odd cardinality respectively.

LEMMA 6.89. *$\widehat{Z}R(x)^2\subseteq\widehat{Z}$; for an arbitrary subset π of $\{1,2,\dots,d\}$ containing at least 2 elements one has: $a_\pi R(x)^2\in a_\pi\widehat{Z}$, $x_\pi R(x)^2\in x_\pi\widehat{Z}$ and $[xv_i]R(x)^2\in[xv_i]\widehat{Z}$.*

PROOF. Let $f\in\widehat{Z}$. Comparing weights we see that $fR(x)^2\in\widehat{Z}$ if d is even and $fR(x)^2=g+a_{12\cdots d}h$; $g,h\in\widehat{Z}$, if d is odd.

If $h\neq 0$ then $fR(x)^2R(v_1)R(x)R(v_2)R(x)\cdots R(v_{d-1})R(x)\notin\widehat{I}$ but $R(x)^2$ commutes with $R(v_1)R(x)R(v_2)\cdots R(v_{d-1})R(x)$ and $fR(v_1)R(x)R(v_2)\cdots R(v_{d-1})R(x)\in\widehat{I}$, a contradiction.

In the proof of Lemma 6.59 it was shown that $a_{ij}R(x)^2\in a_{ij}\hat{Z}$. This implies the assertion for $a_\pi R(x)^2$.

The inclusions $x_\pi R(x)^2\in x_\pi\widehat{Z}x_\tau$ and $[xv_i]R(x)^2\in[xv_i]\widehat{Z}$ follow from Lemma 6.76. Lemma is proved. □

COROLLARY 6.90. $SR(x)^2\subseteq S$.

LEMMA 6.91. *$xv_d\in[xv_d]\widehat{Z}$.*

PROOF. There exists $f\in\widehat{Z}$ such that $([xv_d]f)R(x)=v_d$. Applying $R(x)$ to both sides and using Lemma 6.89 we get the inclusion. Lemma is proved. □

LEMMA 6.92. *(1) For an arbitrary element $a\in S$ if $aR(x)=0$ then $a=0$,*
(2) $\widehat{J}=S+SR(x)$.

PROOF. Let $0\neq f\in\widehat{Z}$, $\pi=\{i_1<\dots<i_{2r}\}$ a nonempty subset of $\{1,2,\dots,d\}$.

It follows from Lemma 6.48 that the only operator $w\in\tilde{W}$ such that $tr((a_\pi f)R(x)w)\neq 0$ is $w=R(v_{i_1})R(x)\cdots R(v_{i_{2r-1}})R(x)R(v_{i_{2r}})$. By Lemmas 6.78 and 6.73 $(a_\pi f)R(x)=x_\pi h$, $0\neq h\in\widehat{Z}$.

Let us show that for an arbitrary $h\in\widehat{Z}$ there exists $f\in\widehat{Z}$ such that this equality holds. Indeed, from the trace form it follows that it is sufficient to choose $f=tr(a_\pi R(x)w)^{-1}tr((x_\pi h)w)$.

Arguing exactly in the same way we conclude that $(x_{i_1\cdots i_{2r+1}}f)R(x) = a_{i_1\cdots i_{2r+1}}h$, where $r \geq 1$, $f, h \in \widehat{Z}$ and $h \neq 0$ as long as $f \neq 0$. Moreover, $(x_{i_1\cdots i_{2r+1}}\widehat{Z})R(x) = a_{i_1\cdots i_{2r+1}}\widehat{Z}$.

Finally, $[xv_i]fR(x) = v_i h$ was claimed in Lemma 6.80. Lemma is proved. □

We know that

$$[[xv_i], x] = \begin{cases} t^n v_1 & \text{if } d = 2,\ v_2^2 = t^n,\ i = 1 \\ v_i & \text{otherwise,} \end{cases}$$

and, moreover, for an arbitrary $f \in \widehat{Z}$ we have

$$[[xv_i]f, x] = \begin{cases} ft^n v_1 & \text{if } d = 2,\ v_2^2 = t^n,\ i = 1 \\ fv_i & \text{otherwise.} \end{cases}$$

LEMMA 6.93. *For arbitrary elements $a, b \in S$ we have $xR(a)R(b) = xR(ab)$.*

PROOF. Since the equality of the lemma can be reformulated as $aD(x, b) = 0$ and since $D(S, S) = (0)$ by Lemma 6.86, we can assume that both a and b are generators of S.

Let us first show that for arbitrary $f, g \in \widehat{Z}$ we have $x(R(f)R(g) - R(fg)) = 0$. By Lemma 6.69 we have $x(R(f)R(g) - R(fg)) \in \widehat{M_I}$. Comparing weights we see that $x(R(f)R(g) - R(fg)) = 0$ if d is even and $x(R(f)R(g) - R(fg)) = x_{12\cdots d}h$, $h \in \widehat{Z}$ if d is odd.

Let $w = R(x)R(v_1)R(x)R(v_2)\cdots R(x)R(v_d)$. If $h \neq 0$ then $tr(x_{12\cdots d}hw) \neq 0$. Let us show that $tr((xR(f)R(g) - xR(fg))w) = 0$. From the assumption that $t^n R(v_1)R(x)\cdots R(v_{d-1})R(x) \in \widehat{I}$ it follows that $(fg)R(x)w \in \widehat{I}$.

We have $tr(((xf)g)w) = tr((xf)gD_1\cdots D_d)$ and
$(xf)gD_1\cdots D_d = \sum \pm(xD_{\mu_1}\cdots D_{\mu_s})R(fD_{\nu_1}\cdots D_{\nu_q})R(gD_{\lambda_1}\cdots D_{\lambda_p})$.

Suppose that $tr((xD_{\mu_1}\cdots D_{\mu_s})R(fD_{\nu_1}\cdots D_{\nu_q})R(gD_{\lambda_1}\cdots D_{\lambda_p})) \neq 0$, $\mu_1, \ldots, \mu_s, \nu_1, \ldots, \nu_q, \lambda_1, \ldots, \lambda_p$ is a permutation of $1, 2, \ldots, d$.

If $0 < p < d$ and d is even, then $gD_{\lambda_1}\cdots D_{\lambda_p} \in \widehat{I} \cup \widehat{M_I}$, which contradicts Lemma 1.28. Let $p = d$ and d is odd. Then $(xf)R(gD_1\cdots D_d) = -gR(v_1)R(x)R(v_2)\cdots R(v_d)R(xf)$. We have $u = gR(v_1)R(x)\cdots R(v_{d-1}) \in \hat{I}$. From the Jordan identity and results of Chapter 1 it follows that $[(ux)v_d, xf] = [(ux)v_d, x]f \bmod \hat{I}$.

Now $[(ux)v_d, x] = gR(v_1)R(x)\cdots R(v_d)R(x) \in x\hat{I}$ by Lemma 6.58. Hence $p = 0$ and therefore $tr((xD_{\mu_1}\cdots D_{\mu_s})R(fD_{\nu_1}\cdots D_{\nu_q})) \neq 0$. For the same reason as above $q = 0$. Since $s = d \geq 2$ it follows that $xD_{\mu_1}\cdots D_{\mu_s} = 0$ because $xD_i = 0$ for $i < d$, the contradiction.

If $a = [xv_i]$, $b = f \in \widehat{Z}$ then the equality $[[xv_i], x]f = [[xv_i]f, x]$ follows from Lemma 6.80.

Now it remains to consider the case $a = [xv_i]$, $b = [xv_j]$.

If $i = j$ then $[x, [xv_i]] \in \widehat{Z}v_i$ and $(\widehat{Z}v_i)[xv_i] = (0)$ by Lemmas 6.59, 6.82.

Suppose that $i \neq j$ and assume that $i < d$. Consider at first the case when $d \geq 3$ or $d = 2$ but $v_2^2 = 1$. Then $[x,[xv_i]][xv_j] = -[xv_j]v_i = xR(a_{ij})R(v_i)R(v_i) = xR(a_{ij}) = x[[xv_i],[xv_j]]$.

Now let $d = 2$, $v_2^2 = t^n$. We have to show that $[x,[xv_2]][xv_1] = x[[xv_2],[xv_1]] = x(a_{21}t^n)$.

Since $[x,[xv_2]] = -v_2$ this amounts to $xR(a_{12})R(v_2)R(v_2) = x(a_{12}t^n)$. Applying the Jordan identity and using Lemmas 6.42 ,6.53 and 6.81 we get the assertion. Lemma is proved. □

Recall that if $\dim V_{n/2}$ is even then we have chosen two isotropic dual subspaces $V_{n/2} = V'_{n/2} + V''_{n/2}$. If $\dim V_{n/2}$ is odd, then we have decomposed $V_{n/2}$ into a direct sum $V_{n/2} = V'_{n/2} + V''_{n/2} + Fw$, where $V'_{n/2}$, $V''_{n/2}$ again are isotropic and dual, $< w, w >= 1$, $< V'_{n/2}, w >=< V''_{n/2}, w >= 0$. Let $v_d = w \otimes t^{n/2}$. Consider the following subspace of $V^{\#}$:

$\tilde{V} = \sum_{-n/2<i<n/2} V_i \otimes t^i + V'_{n/2} \otimes t^{-n/2} + V''_{n/2} \otimes t^{n/2}$.

If $dim_F V_{n/2}$ is even then $\tilde{V} = \sum_{i=1}^{d} Fv_i$. If $dim_F V_{n/2}$ is odd then $\tilde{V} = \sum_{i=1}^{d-1} Fv_i$.

LEMMA 6.94. *S is a graded subspace of $\widehat{J}$.*

PROOF. From what we proved above, Corollary 6.88 and Lemmas 6.48, 6.73 and 6.76, it follows that $S - \{0\}$ consists of elements a such that $a(R(x)R(\tilde{V}))^k R(x) \subseteq \hat{A}$, $a(R(x)R(\tilde{V}))^k R(x) \not\subseteq \widehat{I}$ for some $k \geq 0$, whereas $SR(x)$ consists of elements b such that $b(R(x)R(\tilde{V}))^k R(x) \subseteq \widehat{M} \cup \widehat{I}$ for any $k \geq 0$. Since $\tilde{V}$ is graded it follows that S is graded. Lemma is proved. □

Assume that the element x has been normalized so that either $\deg(x) = 0$ and $t^n R(x)^2 = t^n$ or $deg(x) = -n/2$ (n is even) and $t^n R(x)^2 = 1$.

LEMMA 6.95. *If $v_d^2 = t^n$ then $xv_d = \frac{1}{2}[xv_d]t^{2deg(x)}$.*

PROOF. By Lemma 6.91, $xv_d \in \widehat{Z}[xv_d]$.

The element $[xv_d]$ a priori is not necessarily homogeneous. However, by Lemma 6.94 all homogeneous components of $[xv_d]$ lie in S. From $[[xv_d], x] = v_d$ it follows that for an arbitrary homogeneous component y of $[xv_d]$ of degree other than $\frac{n}{2} - deg(x)$ we have $[y, x] = 0$.

By Lemma 6.92 (1) $y = 0$. Hence, $[xv_d]$ is a homogeneous element and $\deg([xv_d]) = n/2 - deg(x)$. This implies that $xv_d = \alpha[xv_d]t^{2deg(x)}$, where $\alpha \in F$. Applying $R(x)$ to both sides we get $v_d R(x)^2 = \alpha v_d t^{2deg(x)}$. Now, $t^n R(x)^2 = (v_d^2)R(x)^2 = 2v_d(v_d R(x)^2) = 2\alpha t^{n+2deg(x)}$.

Whether $\deg(x) = 0$ or $deg(x) = -n/2$, in both cases $\alpha = 1/2$. Lemma is proved. □

LEMMA 6.96. $(Sx)(Sx) \subseteq S$.

PROOF. Since for arbitrary elements $a, b \in S$ we have $(ax)(bx) = -(-1)^{|b|}(a(bx))x + a(R(x)R(bx) + (-1)^{|b|}R(bx)R(x)) = -(-1)^{|b|}(ab)R(x)^2 + aD(x, bx)$ it follows (see Lemma 6.93 and Corollary 6.90) that it is sufficient to verify the inclusion $(ax)(bx) \in S$ only for a, b from the set of generators $X = \{[xv_i], 1 \leq i \leq d, \widehat{Z}\}$ of S.

Since $\widehat{Z}$ is (topologically) generated by one Laurent variable, it follows that for $a, b \in \widehat{Z}$ we have $(ax)(bx) = (aR(x)^2)b - a(bR(x)^2) \in \widehat{Z}$.

If $a = [xv_i]$, $b = [xv_j]$ then $(ax)(bx) \in (\widehat{Z}v_i)(\widehat{Z}v_j) \subseteq \widehat{Z}$. Now let $a = [xv_i]$, $b = f \in \widehat{Z}$. Again $(ax) \in v_i\widehat{Z}$. Hence $D(ax, f) = (0)$ and therefore $(ax)(fx) = f((ax)x) \in S$ by Corollary 6.90. Lemma is proved. □

Denote $\Gamma = S \cap J$. Then $J = \Gamma + \Gamma R(x)$ and $(\Gamma x)(\Gamma x) \subseteq \Gamma$.

For arbitrary elements $a, b \in \Gamma$ define the bracket $\{a, b\} \in \Gamma$ via $(ax)(bx) = (-1)^{|b|}\{a, b\}$. It is easy to see that $\{b, a\} = -(-1)^{|a||b|}\{a, b\}$.

For an arbitrary element $v = \sum_{i=1}^{d} \alpha_i v_i$, $\alpha_i \in F$, define $[xv] = \sum_{i=1}^{d} \alpha_i [xv_i]$.

For the sake of uniformity even for the case $d = 2$, $v_2^2 = t^n$ we will denote $[xv_1] = xR(a_{12})R(v_2)R(t^{-n})$. Then even in this case $[[xv_1], x] = v_1$.

LEMMA 6.97. *(1)* $[[xv], x] = v$.

(2) If $v = \sum_{i=1}^{d} \alpha_i v_i$, $\alpha_i \in F$ *is homogeneous then* $[xv]$ *is homogeneous as well and* $deg[xv] = deg(v) - deg(x)$.

PROOF. (1) is obvious.

To prove (2) it suffices to repeat the arguments from the proof of Lemma 6.95. Lemma is proved. □

Let us clarify the structure of the superalgebra Γ.

The superalgebra Γ is generated by the even elements t^n, t^{-n} and by the odd graded space $[xV^*]$, where $V^* = \sum_{i=1}^{d} Fv_i$.

Let $G([xV^*])$ be the Grassman algebra on the space $[xV^*]$. Clearly, $\Gamma \simeq F[t^{-n}, t^n] \otimes G([xV^*])$. If we forget about gradation then $\Gamma \simeq F[t^{-1}, t] \otimes G(V)$.

As a graded algebra Γ is isomorphic to the following twisted construction.

Let $deg(x) = 0$. The $\mathbb{Z}/n\mathbb{Z}$-gradation on V induces a $\mathbb{Z}/n\mathbb{Z}$- gradation on $G(V)$. In this case $\Gamma \simeq \mathcal{L}(G(V)) = \sum_{i \in \mathbb{Z}} G(V)_{\bar{i}} \otimes t^i$.

Let $deg(x) = \frac{-n}{2}$. As above, $G(V) = \sum G(V)_i$ is the $\mathbb{Z}/n\mathbb{Z}$-gradation induced from V. In this case (since the gradation on $[xV^*]$ is "shifted" by $\frac{n}{2}$ compared with the gradation on V),

$$\Gamma \simeq \sum_{i \in \mathbb{Z}} G(V)_{\bar{0}\,\bar{i}} \otimes t^i + \sum_{i \in \mathbb{Z}} G(V)_{\bar{1}\,\bar{i}} \otimes t^{i+\frac{n}{2}}.$$

Now we have to define the bracket $\{,\} : \Gamma \times \Gamma \longrightarrow \Gamma$.

Recall that to define a Jordan bracket on Γ we have to define the derivation and all values of the bracket on generators of Γ.

Let us choose a nice basis in V^*. Let $u_1, \dots, u_k$ be a self dual basis in V_0; let $\{u_{i,q}\}_q$ be a basis in $V_i \otimes t^i$, $0 < i < \frac{n}{2}$ and let $\{u_{-i,q}\}_q$ be a dual basis in $V_{-i} \otimes t^{-i}$; let $\{u_{n/2,q}\}$ be a basis in $V' \otimes t^{n/2}$ and let $\{u_{-n/2,q}\}_q$ be a basis in $V''_{n/2} \otimes t^{-n/2}$.

Finally, if $\dim V_{n/2}$ is odd then we add the element $u_d = v_d = w \otimes t^{n/2}$ (recall that $V_{n/2} = V'_{n/2} + V''_{n/2} + Fw$).

Then the algebra Γ is generated by t^n, t^{-n} and $[xu]$, where u runs over all elements in the basis.

Case 1 $\quad t^n R(x)^2 = t^n$, $dim(V_{n/2})$ is even.

In this case $D = \frac{\partial}{\partial t^n} t^n$, $\{t^n, [xu]\} = 0$ for all elements u in the basis,

$$\{[xu'], [xu'']\} = \begin{cases} 0 & \text{if } u'u'' = 0, \\ -1 & \text{if } u'u'' = 1. \end{cases}$$

Case 2 $\quad t^n R(x)^2 = t^n$, $dim(V_{n/2})$ is odd.

We have to add to the previous equalities that
$\{t^n, [xu_d]\} = -\frac{1}{2}[xu_d]t^n$ by Lemma 6.95 and
$\{[xu_d], [xu_d]\} = -t^n$.

Case 3 $\quad t^n R(x)^2 = 1$, $dim(V_{n/2})$ is even.

$D = \frac{\partial}{\partial t^n}$, $\{t^n, [xu]\} = 0$ for all elements u in the basis,

$$\{[xu'], [xu'']\} = \begin{cases} 0 & \text{if } u'u'' = 0, \\ -1 & \text{if } u'u'' = 1. \end{cases}$$

Case 4 $\quad t^n R(x)^2 = 1$, $dim(V_{n/2})$ is odd.

We have to add to Case 3 that $\{t^n, [xu_d]\} = \frac{1}{2}[xu_d]$ and $\{[xu_d], [xu_d]\} = -t^n$.

Let us show that Kantor doubles corresponding to the brackets of cases 1 - 4 are indeed Jordan superalgebras. All of them are embeddable in the Kantor double of Case 3 (of Ramond type) with $n = 1$ and $V = V_0$.

Let $\xi_1, \dots, \xi_{2m}$ be Grassmann variables. Consider the superalgebra of the type K (see [KL]) that corresponds to the differential form $dt - \sum_{i=1}^{m} \xi_i d\xi_{m+i}$ and define a $\mathbb{Z}$-gradation on K via $deg(\xi_1) = 1$, $deg(\xi_{m+1}) = -1$, $deg(t) = deg(\xi_i) = 0$, $i \neq 1, m+1$. This gradation makes the superalgebra $K = K_{-1} + K_0 + K_1$ a Tits-Kantor-Koecher construction of the Jordan superalgebra K_{-1} which is isomorphic to a Kantor double of Case 3.

Now let us embed exceptional superalgebras from Chapter IV (corresponding to the exceptional Cheng-Kac conformal superalgebras CK(6) into the above Kantor doubles. The embedding will not be graded.

We will finish this chapter with embeddings of exceptional superalgebras from chapter IV (corresponding to the exceptional Cheng-Kac conformal superalgebras CK(6)) into Kantor Doubles. The embeddings will be not graded. However that'll be enough to establish that superalgebras defined by tables (T.4.1) and (T.4.2) are Jordan.

In the notation used in this chapter, consider the case $n = 1,\ d = 3, deg(x) = 0,$ $tR(x)^2 = t$. Then the subsuperalgebra generated by $t - \sqrt{-1}a_{123}t,\ v_1,\ v_2,\ v_3,\ x$ is isomorphic to the Jordan superalgebra defined by the table (T.4.1).

The superalgebra defined by the Table (T.4.2) is embeddable into the superalgebra defined by the table (T.4.1). Indeed, if $tR(x)^2 = t$ then $t^2R(xt^{-1})^2 = 2$. Hence, the subalgebra of Table (T.4.1) generated by $t^2,\ v_1,\ v_2\ v_3\ xt^{-1}$ is isomorphic to the superalgebra of Table (T.4.2).

CHAPTER 7

Impossible Cases

In this chapter we consider all cases of the Proposition 1.42 which do not correspond to any Jordan superalgebra satisfying our assumptions

7.1. $I = (0)$, $A = A^{(1)} \oplus A^{(2)}$; $A^{(1)}$ is a loop algebra, $A^{(2)}$ is one-sided graded

PROPOSITION 7.1. Under our assumptions on J the algebra A can not be isomorphic to a direct sum of a loop algebra and a one-sided graded algebra.

PROOF. Let $A^{(1)} \simeq \mathcal{L}(G)$, where $\mathcal{G}$ is a $\mathbb{Z}/n\mathbb{Z}-$graded simple finite dimensional Jordan algebra, and let $A^{(2)}$ be a positively graded algebra.

If e is the identity element of $A^{(1)}$ and f is the identity element of $A^{(2)}$, then from Lemma 5 in [RZ] it follows that $M = \{e, M, f\}$.

We have $A^{(1)} = \oplus_{i=j \, mod \, n} \mathcal{G}_i \otimes t^j$. Let $R(t^n), R(t^{-n})$ denote the multiplications $R(t^{\pm n}) : M \to M$.

From $MU(t^n, t^{-n}) = (0)$ it follows that $2R(t^n)R(t^{-n}) = R(e) = \frac{1}{2}Id$, so $R(t^n)R(t^{-n}) = \frac{1}{4}Id$. This implies that M is a finitely generated module over the subalgebra $< R(t^n), R(t^{-n}) >$ generated by $R(t^n), R(t^{-n})$.

Let $C_+ = \sum_{i>0} Z(A^{(2)})_i$ be the sum of positive components of the center of $A^{(2)}$. We claim that there exists a derivation $d \in \mathcal{D}$ such that $f \in C_+ d$.

Otherwise the subspace C_+ is $\mathcal{D}$-invariant. If $A^{(2)} = \sum_{i \geq -k} A_i^{(2)}$ then $C_+^{k+1} A^{(2)}$ is a $\mathcal{D}$-invariant ideal of $A^{(2)}$ and $C_+^{k+1} A^{(2)} \cap A_0^{(2)} = (0)$. This contradicts the assumption that $I = (0)$.

Now choose a nonzero homogeneous element $c_m \in Z(A^{(2)}) \cap A_m^{(2)}$, $m > 0$ and a derivation $d \in \mathcal{D}$ such that $c_m d = f$.

Consider $M' = \{x \in M |$ there exists $k \geq 1$ such that $xR(c_m^k) = 0\}$.

Since $R(c_m^k)$ commutes with $R(t^n), R(t^{-n})$ we conclude that M' is a finitely generated $< R(t^n), R(t^{-n}) >$-module. Hence there exists $s \geq 1$ such that $M' c_m^s = (0)$.

Suppose that $M' \neq (0)$.

The element c_m^{2s} lies in the annihilator $Ann_{A^{(2)}}(M') = \{a \in A^{(2)} | M'a = (0), AD(M', a) = (0)\}$ (see [Z1]). This annihilator is a $\mathcal{D}$-invariant ideal of $A^{(2)}$.

From $c_m d = f$ it follows that $c_m^{2s} \frac{d^{2s}}{(2s)!} = f$. Hence $f \in Ann_{A^{(2)}}(M')$. But $R(f)$ acts on M as $1/2Id$.

Thus $M' = (0)$ and therefore the operator $R(c_m) : M :\to M$ is injective.

Let $0 \leq k \leq m-1$. Among all integers from $k + m\mathbb{Z}$ choose i_k such that the dimension of M_{i_k} is maximal. Let $l = max\{i_k,\ 0 \leq k \leq m-1\}$. Then for arbitrary $i \geq 0,\ j \geq 0$, we have $M_i = M_{i-mj}c_m^j$ as long as $i - mj \geq l$.

Let us show that $[M, M] \subseteq A^{(1)}$.

For arbitrary elements $x, y \in M,\ a \in A^{(1)}$ we have $[xa, y] - [x, ya] \in A^{(1)}\mathcal{D} \subseteq A^{(1)}$. Hence, $[xa, y]$ and $[x, ya]$ have the same $A^{(2)}$-component.

Choose two homogeneous elements $x, y \in M$. Let $deg(x) + deg(y) = q$. Recall that $A^{(2)} = \sum_{i=-k}^{\infty} A_i^{(2)}$.

Let $\mu \geq 1$ be a sufficiently big number such that $q - m\mu < -k$. For an arbitrary $r \geq 1$ we have $[x, y] = 4[(xt^{nr})t^{-nr}, y]$ and in view of the remark above, to prove that $[x, y] \in A^{(1)}$ it is sufficient to prove that $[xt^{nr}, yt^{-nr}] \in A^{(1)}$.

Choose r big enough so that $deg(x) + nr \geq (\mu + 1)m + l$. Then $xt^{nr} = x'c_m^{\mu+1}$ for some element $x' \in M$.

For arbitrary elements $a \in A^{(2)},\ m', m'' \in M = \{e, M, f\}$ we have $\{m', a, m''\} = a((R(m')R(m'') - R(m'')R(m')) - R([m', m''])) \in \{\{e, M, f\}, \{f, A, f\}, \{e, M, f\}\} \subseteq \{e, A, e\} = A^{(1)}$.

Hence, $aR(m')R(m'') = \frac{1}{2}a(D(m', m'') + R([m', m'']))$ mod $A^{(1)}$.

In particular, $c_m^{\mu+1}R(x')R(yt^{-nr}) = \frac{1}{2}c_m^{\mu+1}(D(x', yt^{-nr}) + R([x', yt^{-nr}])$ mod $A^{(1)}$.

But $deg([x', yt^{-nr}] = q - m(\mu + 1) < -k$, hence $[x', yt^{-nr}] \in A^{(1)}$. And $c_m^{\mu+1}D(x', yt^{-nr}) = (\mu+1)c_m^{\mu}(c_m D(x', yt^{-nr}))$, where the degree of $c_m D(x', yt^{-nr})$ is equal to $s - m\mu < -k$. Since $c_m^{\mu+1}D(x', yt^{-nr}) \in A^{(2)}$ it follows that $c_m D(x', yt^{-nr}) = 0$. The inclusion $[M, M] \subseteq A^{(1)}$ is proved.

Now from $[M, M] \subseteq A^{(1)}$ it follows that $A^{(1)} + M$ is an ideal of J, a contradiction. Proposition is proved. □

7.2. $A = A^{(1)} \oplus A^{(2)}$, $A^{(1)}$ is a negatively graded algebra, $A^{(2)}$ is a positively graded algebra

PROPOSITION 7.2. Under our assumptions on J the algebra A can not be isomorphic to the direct sum of a negatively graded algebra $A^{(1)}$ and a positively graded algebra $A^{(2)}$.

PROOF. As before, let e and f denote the identities of $A^{(1)}$ and $A^{(2)}$ respectively. We showed in section 1 that there exist nonzero central elements $c'_n \in Z(A^{(1)}),\ c''_m \in Z(A^{(2)})$ of degrees $-n < 0,\ m > 0$ respectively and derivations $d', d'' \in \mathcal{D}$ such that $c'_{-n}d' = e,\ c''_m d'' = f$.

Recall that $M = \{e, M, f\}$.

Arguing as in the proof of Proposition 7.1 we can conclude that the multiplications $R(c'_{-n}) : M \to M,\ R(c''_m) : M \to M$ are injective and that M is a finitely generated module over the subalgebra $< R(c'_{-n}), R(c''_m) >$.

Suppose that M is generated by a finite collection of elements $x^{(i)}$ as a $< R(c'_{-n}), R(c''_m) >$-module, $M = \sum_i x^{(i)} < R(c'_{-n}), R(c''_m) >$. Suppose also that $A^{(2)} = \sum_{i=-k}^{\infty} A_i^{(2)}$.

From $\{A^{(1)}, M, A^{(1)}\} = (0)$ it follows that

$$R_M({c'_{-n}}^r)R_M({c'_{-n}}^l) = \frac{1}{2}R_M({c'_{-n}}^{r+l})$$

for arbitrary $r, l \geq 0$ and the similar equality is valid for multiplications by powers of c''_n

Choose a sufficiently big number r such that $nr > 2max\ deg(x^{(i)}) + 2m + k$. We will show that

$$[MR({c'_{-n}}^r), M] \subseteq A^{(1)}.$$

Indeed, we have to prove that

$$a = [x^{(i)}R({c'_{-n}}^{r+l})R({c''_m}^q),\ x^{(j)}R({c'_{-n}}^s)R({c''_m}^\mu)] \in A^{(1)}$$

for arbitrary $l, q, s, \mu \geq 0$.

We will use induction on $q + \mu$.

If $q + \mu \leq 2$, then $deg(a) \leq deg(x^{(i)}) + deg(x^{(j)}) - nr + 2m < -k$ by the choice of r. Hence $a \in A^{(1)}$.

Suppose that $q + \mu \geq 3$. Then $q \geq 2$ or $\mu \geq 2$. Let $q \geq 2$.

Denote $x = x^{(i)}R({c'_{-n}}^{r+l})$, $y = x^{(j)}R({c'_{-n}}^s)R({c''_m}^\mu)$. As in the proof of Proposition 7.1 we can see that:

$$[x{c''_m}^q, y] = \frac{1}{2}{c''_m}^q(D(x,y) + R([x,y])\ mod\ A^{(1)}.$$

The element $[x,y] = [x^{(i)}R({c'_{-n}}^{r+l}), x^{(j)}R({c'_{-n}}^s)R({c''_m}^\mu)]$ lies in $A^{(1)}$ by the induction assumption.

Furthermore, ${c''_m}^q D(x,y) = q{c''_m}^{q-1}(c''_m D(x,y))$ and as above
$c''_m D(x,y) = [x^{(i)}R({c'_{-n}}^{r+l})R(c''_m), x^{(j)}R({c'_{-n}}^s)R({c''_m}^\mu)] -$
$\frac{1}{2}[x^{(i)}R({c'_{-n}}^{r+l}), x^{(j)}R({c'_{-n}}^s)R({c''_m}^{\mu+1})] \in A^{(1)}$ by the induction assumption. The claim is proved.

Similarly there exists $q \geq 1$ such that $[M{c''_m}^q, M] \subseteq A^{(2)}$. Hence,
$[MR({c'_{-n}}^r)R({c''_m}^q), M] \subseteq A^{(1)} \cap A^{(2)} = (0)$.

Now $MR({c'_{-n}}^r)R({c''_m}^q)$ is and ideal of J. This ideal is nonzero since $R({c'_{-n}}^r)$ and $R({c''_m}^q)$ are injective operators, a contradiction. Proposition is proved. □

7.3. $A = A^{(1)} \oplus A^{(2)}$ with $A^{(1)}$ infinite dimensional Jordan algebra of a bilinear form

PROPOSITION 7.3. *A can not be a direct sum $A = A^{(1)} \oplus A^{(2)}$ with one of the summands being an infinite dimensional Jordan algebra of a bilinear form.*

PROOF. Let e and f denote the identity elements of $A^{(1)}$ and $A^{(2)}$ respectively, $A^{(1)} = Fe + \sum_{i \in \mathbb{Z}} V_i$. We know that $M = \{e, M, f\}$ and so $R_M(a)R_M(b) + R_M(b)R_M(a) = R_M(ab)$ for any $a, b \in A^{(1)}$.

If $V_i \neq (0)$, $i \neq 0$, fix elements $v_i \in V_i$, $v_{-i} \in V_{-i}$ such that $< v_i, v_{-i} >= 1$.

Denote $R_i = R_M(v_i)R_M(v_{-i})$. Then $R_i + R_{-i} = R_M(e) = \frac{1}{2}Id_M$.

If $v \in V_j$, $j \neq \pm i$ then $R_M(v)$ anticommutes with $R_M(v_{-i})$ and $R_M(v_i)$ and therefore commutes with R_i.

This implies that $R_iR_j = R_jR_i$ if $i \neq \pm j$.

Remark that $R_M(v_i)^2 = \frac{1}{2}R_M(v_i^2) = 0$. Hence,
$R_i^2 = R_M(v_i)R_M(v_{-i})R_M(v_i)R_M(v_{-i}) =$
$R_M(v_i)(R(v_{-i})R_M(v_i) + R_M(v_i)R_M(v_{-i}))R_M(v_{-i}) =$
$\frac{1}{2}R_M(v_i)R_M(v_{-i}) = \frac{1}{2}R_i$.

This implies that $2R_i$ is an idempotent. Suppose that $VD(M, M) \neq (0)$. The space V is an irreducible module over $D(V, V)$. Since $VD(M, M)$ is $D(V, V)$-invariant it follows that $VD(M, M) = V$.

Choose an element $v \in V$ such that $< v, v > \neq 0$.

Let $v = \sum_\alpha v^{(\alpha)}D(x^{(\alpha)}, y^{(\alpha)})$, where $v^{(\alpha)}$, $x^{(\alpha)}$, $y^{(\alpha)}$ are homogeneous elements from V and M respectively. Since the element v is invertible in $A^{(1)}$ and M is a special $A^{(1)}$-bimodule it follows that $R_M(v)^{-1} = 4R_M(v^{-1})$. Hence
$M = Mv \subseteq \sum_\alpha MD(x^{(\alpha)}, y^{(\alpha)}) + \sum_\alpha(MD(x^{(\alpha)}, y^{(\alpha)}))v^{(\alpha)} \subseteq \sum_\alpha x^{(\alpha)}A + \sum_\alpha y^{(\alpha)}A +$
$\sum_\alpha(x^{(\alpha)}A)v^{(\alpha)} + \sum_\alpha(y^{(\alpha)}A)v^{(\alpha)}$.

Suppose that all elements $x^{(\alpha)}, y^{(\alpha)}$ lie in $M' = \sum_{k=-s}^{s} M_k$, $\dim M' \leq 2sd$. Let $r = 3.2^{2sd}$.

Choose r nonzero numbers $j_1, \ldots, j_r$ such that (1) $V_{j_p} \neq (0)$ for all p, (2) $j_p \neq \pm j_q$ for $p \neq q$, (3) $j_p \neq \pm deg(v^{(\alpha)})$ for all p, α, (4) $M'R_{j_p} \neq (0)$ for all p.

The condition (4) should not be a problem because for any j such that $V_j \neq (0)$ we have $R_j + R_{-j} = \frac{1}{2}Id_M$ and therefore either $M'R_j \neq (0)$ or $M'R_{-j} \neq (0)$.

The operators $R_{j_1}, \ldots, R_{j_r}$ act on M'. Since all of them can be diagonalized simultaneously with only 0 and $\frac{1}{2}$ on the main diagonal there exist 4 distinct numbers k, l, p, q among $j_1, \ldots, j_r$ such that

$$R_k|_{M'} = R_l|_{M'} = R_p|_{M'} = R_q|_{M'}.$$

We claim that $M(R_k - R_l)(R_p - R_q) = (0)$.

Indeed, since k, l, p, q are distinct from $\pm deg(v^{(\alpha)})$ it follows that R_k, R_l, R_p, R_q commute with $R_M(v^{(\alpha)})$. Since the operators R_k, R_l, R_p, R_q lie in $R^M < A^{(1)} >$ it follows that they commute with $R^M < A^{(2)} >$.

It remains to prove that $M'R(v)(R_k - R_l)(R_p - R_q) = (0)$ for any $v \in V$.

If $v \in V_j$ then $\{-j, j\} \cap \{\pm k, \pm l\} = \emptyset$ or $\{-j, j\} \cap \{\pm p, \pm q\} = \emptyset$.

Without loss of generality we can assume the first case. Then $R(v)$ commutes with R_k and R_l and $M'R(v)(R_k - R_l) = M'(R_k - R_l)R(v) = (0)$.

Now $R(v_l)R_l = 0$, $R_qR(v_{-q}) = 0$. Hence $MR(v_l)(R_k - R_l)(R_p - R_q)R(v_{-q}) = MR(v_l)R_kR_pR(v_{-q}) = (0)$. This implies $MR_lR_kR_pR_q = (0)$ and therefore $M'{R_l}^4 = M'R_l = (0)$, which contradicts our choice of $j_1, \ldots, j_r$.

We proved that $VD(M, M) = (0)$. Hence, for arbitrary elements $a \in A^{(1)}$; $x, y \in M$ we have $[xa, y] = [x, ya]$.

Choose an element $v \in V$ such that $< v, v >= 1$. Then $e_1 = \frac{1}{2}(e + v)$, $e_2 = \frac{1}{2}(e - v)$ are orthogonal idempotents in $A^{(1)}$, $e_1 + e_2 = e$. Let $M_i = \{e_i, M, f\}$, $i = 1, 2$; $M = M_1 + M_2$. We have, $[M_i, M_i] \subseteq Fe_i + A^{(2)}$, $[M_1, M_2] = [e_1M_1, M_2] = [M_1, e_1M_2] = (0)$.

Hence, $[M,M] \subseteq Fe_1 + Fe_2 + A^{(2)} = Fe + Fv + A^{(2)}$. Varying v we get $[M,M] \subseteq Fe + A^{(2)}$. This implies that $[M_i, M_i] \subseteq (Fe_i + A^{(2)}) \cap (Fe + A^{(2)}) = A^{(2)}$ and therefore $[M,M] \subseteq A^{(2)}$. Now $A^{(2)} + M$ is an ideal of J, a contradiction. Proposition is proved. $\square$

7.4. $I \neq (0)$, A/I is an infinite dimensional Jordan algebra of a nondegenerate symmetric bilinear form

PROPOSITION 7.4. Under our assumptions on J the above case is impossible.

Let W be a finite dimensional vector space over F with a symmetric bilinear nondegenerate form $<,>: W \times W \to F$. We will recall the classification of Jordan bimodules over the Jordan algebra $B = F1 + W$ (see [J3]).

Consider the Clifford algebra $Cl(W)$. The Jordan algebra $B = F1 + W$ is a subalgebra of $Cl(W)^{(+)}$. Hence $Cl(W)$ is a Jordan bimodule over B. For k elements $v_1, \dots, v_k \in W$ the k-vector $[v_1, \dots, v_k] \in Cl(W)$ is defined inductively: $[v_1] = v_1$; if $k \geq 2$ is even then $[v_1, \dots, v_k] = [[v_1, \dots, v_{k-1}], v_k]$, where $[a,b] = ab - ba$ is the commutator in $Cl(W)$; if $k \geq 3$ is odd then $[v_1, \dots, v_k] = [v_1, \dots, v_{k-1}] \cdot v_k = \frac{1}{2}([v_1, \dots, v_{k-1}]v_k + v_k[v_1, \dots, v_{k-1}])$. It is not difficult to see that $[v_{\sigma(1)}, \dots, v_{\sigma(k)}] = (-1)^{|\sigma|}[v_1, \dots, v_k]$ for an arbitrary permutation σ.

Let C_k be the linear span of all k-vectors, $C_0 = F1$.

THEOREM 7.5. *([J3]) Suppose that the dimension $dim_F W = 2q$ is even.*
1) For any integer $0 \leq k \leq q$, $M(k) = C_{2k} + C_{2k+1}$ is an irreducible B-subbimodule of $Cl(W)$. If $k = q$ then $M(q) = C_{2q}$ is a one-dimensional bimodule.

2) An arbitrary unital B-bimodule is isomorphic to a direct sum of $M(k)'s$, $0 \leq k \leq q$.

Now let the vector space $W = \sum_{i \in \mathbb{Z}} W_i$ be $\mathbb{Z}$-graded. For an arbitrary integer $l \in \mathbb{Z}$ we will define a gradation on $M(k)$: If $v_1, \dots v_r$ are homogeneous elements, $r = 2k$ or $r = 2k+1$, then $deg[v_1 \dots, v_r] = deg(v_1) + \cdots + deg(v_r) + l$. Let us denote as $M(k,l)$ the above defined graded B-bimodule.

From Theorem 7.5 it follows that an arbitrary graded B-bimodule is isomorphic to a direct sum of $M(k,l)'s$, $0 \leq k \leq q$, $l \in \mathbb{Z}$. Indeed, let S be a graded B-bimodule. Theorem 7.5 implies that the multiplication algebra $R^S < B >$ is a direct sum of matrix algebras over F, $R^S < B > \simeq \oplus_k M_{n_k}(F)$. The $\mathbb{Z}$-gradation on B induces a $\mathbb{Z}$-gradation on $R^S < B > = \sum_{i \in \mathbb{Z}} R^S < B >_i$. Consider the semisimple derivation $d : x \to ix$, $x \in R^S < B >_i$. Since all ideals $M_{n_k}(F)$ are d-invariant it follows that $M_{n_k}(F)'s$ are graded ideals. Moreover, it is easy to see that $M_{n_k}(F) \cap R^S < B >_0$ contains n_k primitive orthogonal idempotents. Hence $R < B > = \sum_j \rho_j$, where the $\rho_j's$ are minimal right ideals which are graded. This implies that S is a direct sum of bimodules $M(k)$ with some gradation on them.

Let us show that the only $\mathbb{Z}$-gradations on $M(k)$ are those of type $M(k,l)$.

Choose arbitrary linearly independent homogeneous elements $v_1, \dots, v_{2k} \in W$. The elements $0 \neq x = [v_1, \dots, v_{2k}] \in M(k)$ generates $M(k)$ as a B-bimodule. Hence a degree of x uniquely determines a gradation on $M(k)$. If $l = deg(x) - \sum_{i=1}^{2k} deg(v_i)$ then the gradation is of the type $M(k, l)$.

LEMMA 7.6. *If* $0 < k < q$ *then* $dim_F M(k, l)_l \geq q$.

PROOF. Choose a basis $v_{\alpha_1}, v_{-\alpha_1}, \dots, v_{-\alpha_q}, v_{\alpha_q}$ of W consisting of homogeneous elements of degrees $-\alpha_1, \alpha_1, \dots, -\alpha_q, \alpha_q$ respectively, such that $< v_{\alpha_i}, v_{\alpha_i} >= 1$, $1 \leq i \leq q$, and all other products are equal to 0.

For an arbitrary subset $\{i_1, \dots, i_k\}$ of $\{1, 2, \dots, q\}$ the $2k$-vector $[v_{-\alpha_1}, v_{\alpha_1}, \dots v_{-\alpha_k}, v_{\alpha_k}]$ lies in $M(k, l)_l$ and, moreover, for distinct subsets these vectors are linearly independent. Hence $dim_F M(k, l)_l \geq q$. Lemma is proved. □

Remark that $M(0)$ is isomorphic to the regular bimodule and that $dim_F M(q) = 1$.

PROOF OF PROPOSITION 7.4

Suppose that $I = M(A) \neq (0)$, $A/I \equiv F1 + \bar{V}$ is the Jordan algebra of a symmetric bilinear nondegenerate form in an infinite dimensional vector space $\bar{V}$. Let $^- : A \to F1 + \bar{V}$ be the natural homomorphism.

As we have seen in Chapter 1 there exist elements $x, y \in M$; $u \in I$; $a, b \in A$ such that $t([(xu)a, y]b) \neq 0$.

Without loss of generality we can assume that $\bar{a} \in \bar{V}_i$, $\bar{b} \in \bar{V}_j$. Moreover, the degrees i, j can be assumed to be nonzero. Indeed, let $i = 0$. We can always assume that $< \bar{a}, \bar{b} >= 0$. Choose an integer k such that $k \neq 0$, $k \neq j$ and $\bar{V}_k \neq (0)$. Choose elements $\bar{c}_{-k} \in \bar{V}_{-k}$, $\bar{c}_k \in \bar{V}_k$ such that $< \bar{c}_{-k}, \bar{c}_k >= 1$, $< \bar{c}_k, \bar{a} >=< \bar{c}_{-k}, \bar{a} >=< \bar{c}_k, \bar{b} >=< \bar{c}_{-k}, \bar{b} >= 0$. Let c_k be a preimage of $\bar{c}_k$. We have $\bar{a} = \bar{c}_k D(\bar{c}_{-k}, \bar{a})$. Since $\bar{b} D(\bar{c}_{-k}, \bar{a}) = 0$ it follows that $t([(MI)c_k, M]b) \neq (0)$.

From $ij \neq 0$ it follows that $\bar{a}$, $\bar{b} \in \bar{V}_{-r} + \dots + \bar{V}_{-1} + \bar{V}_1 + \dots + \bar{V}_r = \bar{V}'$, where $r > max(3, d)$.

We will need a Wedderburn splitting theorem for graded Jordan algebras to claim the existence of a graded subspace $V' \subseteq A$ such that $^- : F1 + V' \to F1 + \bar{V}'$ is an isomorphism.

In [J3, chapter 6, sections 6-7] the following idea is used to prove a splitting theorem for Jordan algebras. Let J be a finite dimensional Jordan algebra, $L = K(J)$ its Tits-Kantor-Koecher Lie algebra. There is an involutive automorphism $\epsilon L \to L$ such that subalgebras of J correspond to ϵ-invariant subalgebras of L. E. Taft [Ta] proved that for an arbitrary finite group of automorphisms $G \leq Aut(L)$ there exists a G-invariant semisimple subalgebra S of L such that $L = Rad(L) + S$. Applying this theorem to the group $< \epsilon >$ we get the splitting theorem for J.

Now suppose that $J = \sum_{i=-n}^{n} J_i$ is $\mathbb{Z}$-graded. This gradation gives rise to a a $\mathbb{Z}$-gradation $L = \sum_{i=-2n}^{2n} L_i$. Let $\eta_1, \eta_2, \dots, \eta_{2n}, \eta_{2n+1} = 1$ be the distinct $2n+1$-roots of 1 and define $2n$ automorphisms of L by $\varphi_i : a_j \to \eta_i^j a_j$, $1 \leq i \leq 2n$, $j \in \mathbb{Z}$, $a_j \in L_j$. The subgroup G of $Aut(L)$ generated by the automorphisms $\epsilon, \varphi_1, \dots, \varphi_{2n}$ is abelian and finite. By Taft's theorem there exists a G-invariant semisimple subalgebra $S \subseteq L$ such that $L = Rad(L) + S$. Since S is ϵ-invariant it follows that S corresponds to a semisimple subalgebra $J' \leq J$ such that $J = Rad(J) + J'$. Since J' is $< \varphi_1, \dots, \varphi_{2n} >$- invariant it follows that J' is graded. We proved a graded version of the splitting theorem.

Thus there exists a graded subspace $V' \subseteq A$ such that $^- : F1 + V' \to F1 + \bar{V}'$ is an isomorphism, $dim_F V'$ is even, $dim_F V' > d$.

As we did in chapters 4 and 6 choose $n = dim_F V' \geq 6$ nonhomogeneous involutions $v_1, \dots, v_n \in V'$ which span V', $v_i v_j = \delta_{ij}$.

There exist integers $1 \leq i \neq j \leq n$ such that $t([(MI)v_i, M]v_j) \neq (0)$. Renumerating involutions we can assume that $t([(MI)v_1, M]v_2) \neq (0)$.

From Lemma 7.6 it follows that in decompositions of I and M into direct sums of graded irreducible $B = F1 + V'$-modules only regular bimodules and 1-dimensional bimodules can occur. Let $I = (\oplus_\alpha \psi_\alpha(B)) \oplus I'$; $M = (\oplus_\beta \varphi_\beta(B)) \oplus M'$, where $\psi_\alpha : B \to I$, $\varphi_\beta : B \to M$ are B-bimodule isomorphisms; I' and M' are sums of 1-dimensional B-bimodules. Denote

$I_1 = \sum_\alpha F\psi_\alpha(1)$, $I_{v_i} = \sum_\alpha F\psi_\alpha(v_i)$,
$M_1 = \sum_\beta F\varphi_\beta(1)$, $M_{v_i} = \sum_\beta F\varphi_\beta(v_i)$.

Then $M = M_1 + \sum_{i=1}^n M_{v_i} + M'$, $I = I_1 + \sum_{i=1}^n I_{v_i} + I'$.

In view of our assumption there exist $x, y \in M_1 \cup M_{v_1} \cup \cdots M_{v_n} \cup M'$, $u \in I_1 \cup I_{v_1} \cup \cdots \cup I_{v_n} \cup I'$ such that $t([(xu)v_1, y]v_2) \neq 0$.

If $xv_3 = uv_3 = yv_3 = 0$ then $([(xu)v_1, y]v_2)v_3 = 0$ which contradicts our assumption. Hence, at least one of the elements x, u, y lies in $M_1 \cup M_{v_3}$ or in $I_1 \cup I_{v_3}$. Similarly, for each $k = 4, 5, 6$ at least one of the elements x, u, y lies in $M_1 \cup M_{v_k}$ or in $I_1 \cup I_{v_k}$. This implies that $x \in M_1$ or $y \in M_1$ or $u \in I_1$.

Let $y \in M_1$. Since y belongs to the eigevalue 1 with respect to $U(v_i)$ it follows that $R(v_1)R(y) = R(y)R(v_1)$. Hence, $[(xu)v_1, y] = [xu, y]v_1 \in I$ by Lemma 1.27, a contradiction. The cases $x \in M_1$, $u \in I_1$ are treated similarly. Proposition is proved.

7.5. $A = A^{(1)} \oplus A^{(2)}$, $A^{(1)}$ is finite dimensional; $A^{(2)}$ is a loop algebra

PROPOSITION 7.7. Under our assumption on J the algebra A can not be isomorphic to a direct sum $A^{(1)} \oplus A^{(2)}$, where $A^{(1)}$ is a finite dimensional simple algebra and $A^{(2)}$ is isomorphic to a loop algebra $\mathcal{L}(G)$ of a $\mathbb{Z}/n\mathbb{Z}$-graded finite dimensional simple algebra $\mathcal{G}$.

Assume the contrary. Let e' and e'' be the identities of $A^{(1)}$ and $A^{(2)}$ respectively. From [R-Z] it follows that $M = \{e', M, e''\}$. Let $Z(A^{(2)})$ be the center of $A^{(2)}$.

LEMMA 7.8. $Z(A^{(2)})D(M, M) = (0)$.

PROOF. Unless $A^{(2)} \simeq F[t^{-n}, t^n]$ the identity element e'' is a sum of two orthogonal(not necessarily homogeneous) idempotents: $e'' = e_1'' + e_2''$. Then $M =$

$\{e', M, e''_1\} + \{e', M, e''_2\}$ and therefore

$$Z(A^{(2)})D(\{e', M, e''_1\}, \{e', M, e''_2\}) \subseteq Z(A^{(2)}) \cap \{e''_1, A^{(2)}, e''_2\} = (0),$$

$$Z(A^{(2)})D(\{e', M, e''_i\}, \{e', M, e''_i\}) \subseteq Z(A^{(2)}) \cap \{e''_i, A^{(2)}, e''_i\} = (0).$$

Hence we will assume that $A^{(2)} \simeq F[t^{-n}, t^n]$.

Denote $c = t^n$. If $cD(M, M) \neq (0)$ then there exists a (not necessarily homogeneous) element $x \in M$ such that $cR(x)^2 \neq 0$.

Let us show that for an arbitrary $k \geq 1$ we have $cR(xc^k)^2 \neq 0$.

Indeed, it is easy to check that for an arbitrary element $y \in M$ and arbitrary integers i, j one has $[yc^i, yc^j] = \frac{i-j}{2} c^{i+j-1}[yc, y]$.

This implies that $cR(xc^k)^2 = [c(xc^k), xc^k] = \frac{1}{2}[xc^{k+1}, xc^k] = \frac{1}{4}c^{2k}[xc, x] \neq 0$.

Thus $cR(xc^k)^2 = f_k(c)$ is a nonzero Laurent polynomial.

Consider the completion $\widehat{J} = \widehat{A} + \widehat{M}$ of the superalgebra J.

For an arbitrary homogeneous element $z \in M$ we have

$$z = 4(zf_k(c))f_k(c)^{-1} = 4(z(cR(xc^k)^2))f_k(c)^{-1} =$$

$$4([zc, xc^k](xc^k) - (([z, xc^k](xc^k))c))f_k(c)^{-1}.$$

If k is sufficiently large, then $[zc, xc^k]$ and $[z, xc^k]$ can lie only in $A^{(2)}$. Hence $z \in x\widehat{A^{(2)}}$ and therefore $\widehat{M} = x\widehat{A^{(2)}}$. Now it follows that $[\widehat{M}, \widehat{M}] = [x\widehat{A^{(2)}}, x\widehat{A^{(2)}}] \subseteq [(x\widehat{A^{(2)}})\widehat{A^{(2)}}, x] + \widehat{A^{(2)}}D(x\widehat{A^{(2)}}, x) \subseteq \widehat{A^{(2)}}R(x)^2 + \widehat{A^{(2)}}D(x\widehat{A^{(2)}}, x) \subseteq \widehat{A^{(2)}}$.

This contradicts simplicity of J, because $A^{(2)} + M$ is now a proper ideal of J. Lemma is proved. □

LEMMA 7.9. *(1)* $D(A, M) = D(A^{(2)}, M)$,

(2) $MV(A^{(2)}, M) \subseteq A^{(1)}$,

(3) $MV(A^{(2)}, M)V(A^{(2)}, M) = (0)$,

(4) $AV(A^{(2)}, M)V(A^{(2)}, M)V(A^{(2)}, M) = (0)$.

PROOF. Let $a \in A^{(1)}$, $x \in M$. We have $x = x'c$, $c \in A^{(2)}$.

Hence, $D(a, x'c) = D(x'a, c) + D(ac, x') = D(x'a, c) \in D(M, A^{(2)})$.

The assertions (2)-(4) follow from Peirce relations. Lemma is proved. □

Let S be the subalgebra of the multiplication algebra $R < J >$ generated by all multiplications $R(c^i)$, $i \in \mathbb{Z}$, by $\mathcal{D}(M, M)$ and by $\mathcal{D}(A, A)$.

Since M is a free module of rank $\leq dn$ over $< R_M(c), R_M(c^{-1}) >$ (indeed, M is a graded module over the algebra of Laurent polynomials) and $S|_M$ lies in the endomorphism ring of this module, we conclude that $S|_M$ is embedable into a matrix ring over $< R_M(c), R_M(c^{-1}) >$.

Similarly $S|_{A^{(1)}}$ is finite dimensional and $S|_{A^{(2)}}$ is embeddable into a matrix ring over $< R(c)|_{A^{(2)}}, R(c^{-1})|_{A^{(2)}} >$.

This implies that S is a finite right module over the subalgebra $< R(c), R(c^{-1}) >$.

Choose two homogeneous elements $x, y \in M$. We have $x = x'c^i$, $y = y'c^j$, where $0 \leq deg(x') < n$, $0 \leq deg(y') < n$.

Since $D(c^j, A) = (0)$, for an arbitrary element $z \in M$ we have

$$\begin{aligned} zR(x)R(y) = [z,x](y'c^j) = c^j(y'[z,x]) = \\ zR(x)R(y')R(c^j) = -zR(y')R(x)R(c^j) + \\ zD(x,y')R(c^j) = -zR(y')R(x')R(c^i)R(c^j) + zD(x,y')R(c^j). \end{aligned} \tag{7.1}$$

If $a \in A^{(2)}$ is a homogeneous element and $a = a'c^l$, $0 \leq deg(a') < n$, then

$$R_M(a) = 2R_M(a')R_M(c^l) \tag{7.2}$$

Now consider the operator $R(x)R(a)R(y)$; $x = x'c^i$, $y = y'c^j$, $a \in A$. For an arbitrary element $z \in M$ we have $zR(x)R(a)R(y) =$
$zR(x)R(a)R(y')R(c^j)$. Furthermore

$$zR(x)R(a)R(y') = zR(x)R(y')R(a) + zR(x)D(a,y'). \tag{7.3}$$

The first summand was considered in (7.1). As for the second one,
$zR(x)D(a,y') = zR(xD(a,y')) - zD(a,y')R(x) =$

$$zR(xD(a,y')) - zD(a,y')R(x')R(c^i) \tag{7.4}$$

Finally, we have to assume that $a \in A^{(2)}$, $a = a'c^k$, $0 \leq deg(a') < n$ and to take care of the summand $zR(y')R(a'c^k)R(x') = zR(y')R(a')R(c^k)R(x')$. Moreover, since $\{zR(y')R(a'), x', c^k\} \in \{A^{(2)}, M, A^{(2)}\} = (0)$ it follows that

$$zR(y')R(a')R(c^k)R(x') = 2zR(y')R(a')R(x')R(c^k). \tag{7.5}$$

LEMMA 7.10. *$R < J >= (R(J)R(J) + R(J)R(J)D(M,A) +$
$R(J)R(J)D(M,A)D(M,A))S$.*

PROOF. Using formula (D) from the Introduction we see that an arbitrary operator v is a sum of operators of the type $v'D_1 \cdots D_k$, where v' is an operator of length ≤ 2, $D_i \in D(A,A) \cup D(M,M) \cup D(M,A)$.

If at least one D_i lies in $D(A,A) \cup D(M,M)$ then we can move it to the right end modulo shorter operators.

If all D_i lie in $D(M,A)$ we substitue $D(a_i,x_i) = V(a_i,x_i) - R(a_ix_i)$ and then use parts (3),(4) of the previous lemma. Lemma is proved. □

The multiplication algebra $R = R < J >$ is $\mathbb{Z}/2\mathbb{Z}$-graded, $R = R_{\bar{0}} + R_{\bar{1}}$. The even part $R_{\bar{0}}$ consists of those operators that map M to M and A to A, whereas odd operators (from $R_{\bar{1}}$) map M to A and A to M.

From the previous lemma it follows that

$$\begin{aligned} R_{\bar{0}} = (R(A)R(A) + R(M)R(M) + R(A)R(M)D(M,A) + \\ R(M)R(A)D(M,A) + R(A)D(M,A)D(M,A) + \\ R(A)R(A)D(M,A)D(M,A) + R(M)R(M)D(M,A)D(M,A))S \end{aligned} \tag{7.6}$$

Using $D(M,A) \subseteq V(A^{(2)},M) + R(M)$ and part (3) of the previous lemma we get $R_{\bar{0}}|_M = (R(A)R(A) + R(M)R(M) + R(A)R(M)D(M,A))|_M S|_M$.

Now (7.1) - (7.5) imply that $R_{\bar{0}}|_M$ is a finite right module over $S|_M$ and hence over $< R(c)|_M, R(c^{-1})|_M >$. Since $R_{\bar{0}}|_M$ is a graded module over the Laurent polynomial algebra $< R(c)|_M, R(c^{-1})|_M >$ we conclude that $R_{\bar{0}}|_M$ is a free right $< R(c)|_M, R(c^{-1})|_M >$-module, say, of rank r. Now $a \to L(a)$, where $a \in R_{\bar{0}}|_M$, $L(a)$ is a left multiplication by a, is an antiisomorphism of $R_{\bar{0}}|_M$ into the algebra of $r \times r$ matrices over $< R(c)|_M, R(c^{-1})|_M >$. We have proved the following

LEMMA 7.11. *The associative algebra $R_{\bar{0}}|_M$ is PI.*

Let $\tilde{S}$ be the subalgebra of R generated by $D(A,A)$, $D(M,M)$, $R(A)$, $U(M,M)$. Let $\tilde{S}_0$ be the subalgebra generated by $D(A,A)$, $D(M,M)$, $R(A)$. Let $\tilde{S}_1 = U(M,M)\tilde{S}_0$. By McDonald's theorem (see [J3], [ZSSS]) the following identities hold in all Jordan algebras

$$R(x)U(y) + U(y)R(x) = 2U(y, yx),$$
$$U(x)U(y) = 2v(x,y)^2 - V(x, xU(y)).$$

The first of these identities implies that $\tilde{S}_1 = \tilde{S}_0 U(M,M)$, hence $\tilde{S}_0\tilde{S}_1 \subseteq \tilde{S}_1$, $\tilde{S}_1\tilde{S}_0 \subseteq \tilde{S}_1$. The second identity implies that $\tilde{S}_1\tilde{S}_1 \subseteq \tilde{S}_0$. Hence $\tilde{S} = \tilde{S}_0 + \tilde{S}_1$, $\tilde{S}_i\tilde{S}_j \subseteq \tilde{S}_{i+j\, mod\, 2}$. The intersection $Q = \tilde{S}_0 \cap \tilde{S}_1$ is an ideal of $\tilde{S}$. The quotient algebra $\tilde{S}/Q$ is an associative superalgebra whose even part is PI. By the theorem of V. K. Kharchenko [Kh] the algebra $\tilde{S}/Q$ is PI. Since the algebra $Q \subseteq \tilde{S}_0$ is PI as well we conclude that the algebra $\tilde{S}$ is PI.

Let $L = R_{\bar{1}}V(A^{(2)},M)$ be the right ideal of $R_{\bar{0}}$. From (7.6), Lemma 7.9(1) and the inclusion $D(A^{(2)},M) \subseteq V(A^{(2)},M) + R(M)$ it follows that $R_{\bar{0}} = L + \tilde{S}$

The operators from $L|_A$ map A to $A^{(1)}$. Hence, the algebra $L|_A$ is PI. Now we will need a lemma from the theory of PI-algebras

LEMMA 7.12. *(see [R2]) Let R be an associative algebra, L a left ideal of R and B a subalgebra of R. Suppose that $R = L + B$ and both algebras L and B are PI. Then the algebra R is PI.*

PROOF. Clearly, $I = L + LB$ is a two sided ideal of R. If L satisfies a multilinear identity $f(x_1, x_2, \ldots) = 0$, then for an arbitrary $b \in B$ the left ideal Lb satisfies the identity $x_0 f(x_1, x_2, \ldots) = 0$. Hence, I is a sum of left ideals which are PI. L. Rowen proved (see [R2, Ex. 7.2.2]) that a sum of two left ideals which ate PI is a PI-algebra. Hence I is a locally PI-algebra, that is, every finitely generated subalgebra of I is PI. The quotient algebra $R/I \simeq B/B \cap I$ is a PI-algebra as well.

Suppose that the algebra R is not PI. For an arbitrary nonzero element $0 \neq f(x,y)$ of the free associative algebra on two generators x, y choose elements $a_f, b_f \in R$ such that $f(a_f, b_f) \neq 0$. Let $\{R_f\}_f$ be a collection of isomorphic copies of R indexed by nonzero elements of the free associative algebra. Let $\tilde{R} = \prod_f R_f$ be the Cartesian product, $\tilde{a} = \prod_f a_f$, $\tilde{=} \prod_f b_f$. then the elements $\tilde{a}$, $\tilde{b}$ generate the free associative algebra (this argument is known as the Amitsur trick [R2]). On the other hand, $\tilde{R} = \prod_f L_f + \prod_f B_f$. Hence $\tilde{R}$ is an extension of a locally PI-algebra by a PI-algebra, and thus can not contain a free subalgebra, the contradiction. Lemma is proved. □

From this lemma it follows that the algebra $R_{\bar{0}}|_A$ is PI. Together with Lemma 7.11 this implies that the algebra $R_{\bar{0}}$ is PI and moreover (see [R2]) the algebra R is PI.

PROOF OF PROPOSITION 7.7 Suppose that the Jordan superalgebra $J = A+M$ satisfies the assumptions of the proposition, $A = A^{(1)} \oplus A^{(2)}$, where $\dim A^{(1)} < \infty$, $A^{(2)}$ is a loop algebra. As we have shown above the multiplication algebra $R = R < J >$ is PI. Since the superalgebra J is graded simple, it is a graded irreducible faithful module over R. This implies that R is graded prime and hence prime. Since $A^{(2)}$ is a loop algebra, the graded algebra R has infinitely many nonzero homogeneous components. By Lemma 3.5 there exists a homogeneous operator $w \in R$ of degree $i \neq 0$ lying in the center of R.

As in chapter 3 we can assume that $w \in R_{\bar{0}}$.

Notice that $A^{(1)}w \subseteq A^{(1)}$. Indeed, $A^{(1)} = A^{(1)}A^{(1)}$. Hence, $A^{(1)}w = (A^{(1)}A^{(1)})w \subseteq (A^{(1)}w)A^{(1)} \subseteq AA^{(1)} = A^{(1)}$. Since $\dim A^{(1)} < \infty$ and $\deg(w) = i \neq 0$, we conclude that $A^{(1)} \cap ker(w) \neq (0)$. But $ker(w)$ is an ideal of the superalgebra J, hence $ker(w) = (0)$, the contradiction. Proposition is proved. □

Acknowledgment. The authors are grateful to a referee who carefully read the paper and made many valuable suggestions.

Bibliography

[AK] A. D'Andrea and V.G. Kac, *Structure theory of finite conformal algebras*, Selecta Math., to appear.

[B] Yu. A. Bakhturin, *Lectures on Lie Algebras*, Akademie-Verlag, Berlin, 1978.

[CK] S.J. Cheng and V.G. Kac , *A new N=6 superconformal algebra*, Commun. Math. Phys. 186, 219-231, 1997.

[J1] N. Jacobson, *Structure of Rings*, Amer. Math. Soc. Coll. Publ. 37, 1956.

[J2] N. Jacobson, *Lie algebras*, Interscience, New York, 1962.

[J3] N. Jacobson, *Structure and representation of Jordan Algebras*, Amer. Math. Soc. Providence, R.I., 1969.

[K1] V. Kac, *Simple graded Lie algebras of finite growth*, Math. USSR Izv 2, 1271-1311, 1968.

[K2] V. Kac, *Lie superalgebras*, Advances in Math. 26, 8-96, 1977.

[K3] V. Kac, *Classification of simple Z-graded Lie superalgebras and simple Jordan superalgebras*, Comm. in Algebra 5, 1375-1400, 1977.

[K4] V. Kac, *Infinite dimensional Lie algebras*, Cambridge University Press,1990.

[K5] V. Kac, *The idea of locality*,in Physical applications of geometry, groups and algebras, H.D. Doebner et al. editors, World Sci., Singapore, pp 16-22, 1997.

[K6] V. Kac, *Superconformal algebras and transitive group actions on quadrics*, Comm. Math. Phys. 186, 233-252, 1997.

[K7] V. Kac, *Classification of infinite-dimensional linearly compact simple Lie superalgebras*, to appear.

[KL] V.G. Kac and J.W. van de Leur, *On classification of superconformal algebras*, Strings 88, World Sci., 77-106, 1989.

[Ka1] I.L. Kantor, *Transitive differential groups and invariant connections in homogeneous spaces*, Trudy sem. vect. tensor. anal. 13, 310-398, 1966.

[Ka2] I.L.Kantor, *Jordan and Lie superalgebras defined by Poisson algebras*, preprint, 1991.

[Kap] I. Kaplansky, *Superalgebras*, Pacific J. Math. 86, 93-98, 1980.

[KM] D. King and K. McCrimmon, *The Kantor Construction of Jordan Superalgebras*, Comm. in Algebra 20(1), 109-126, 1992.

[Kh] V. K. Kharchenko, *Galois extensions and rings of fractions*, Algebra and Logic 13, 265-281, 1974.

[Ki] A. A. Kirillov, *Local Lie Algebras*, Russian Math. Surveys 31:4, 55-76, 1976.

[Ko] M. Koecher, *Imbedding of Jordan algebras into Lie algebras*, I. Amer. J. Math. 89, 787-816, 1967.

[NS] A. Neveu and J.H. Schwarz, *Factorizable dual models of pions*, Nucl. Phys. B 31, 86-112, 1971.

[M] V.T. Markov, *On the dimension of noncommutative affine algebras*, Math. USSR, Izv. 7 (1973).

[MZ1] C. Martinez and E. Zelmanov, *Jordan algebras of GK-dimension 1*, J. of Algebra 180, 211-238, 1996.

[MZ2] . Martinez and E. Zelmanov, *Simple and Prime graded Jordan algebras*, J. of Algebra 196, 596-613, 1997.

[Ma1] O. Mathieu, *Classification des algebres de Lie graduees simples de croissance* ≤ 1, Invent. Math. 86, 371-326, 1986.

[Ma2] O. Mathieu, *Classification of Simple Graded Lie Algebras of Finite Growth*, Invent. Math. 108, 455-519, 1992.

[RZ] M. Racine and E. Zelmanov, *Classification of simple Jordan superalgebras with semisimple even part*, Preprint, 1998.

[Ra] P. Ramond, *Dual theory for free fermions*, Phys. Rev. D3, 2415-2418, 1971.

[Re] A. Regev, *Existence of identities in $A \otimes B$*, Israel J. Math. 11, 131-152, 1972.

[R1] L Rowen, *Some results on the center of a ring with polynomial identity*, Bull. AMS 79, 219-223, 1973.

[R2] L Rowen, *Polynomial Identities in Ring Theory*, Academic Press, New York, 1985

[R3] L Rowen, *Ring Theory*, Academic Press, New York, 1991.

[Sh] I.P. Shestakov, *Finite dimensional algebras with a nil basis*, Algebra i Logika 10, N.1, 87-99, 1971.

[S] V.G. Skosirskii, *On nilpotence in Jordan and right alternative Algebras*, Algebra i Logika 18 n.1, 73-85, 1979.

[Ta] E.J. Taft, *Invariant splitting in Jordan and alternative algebras*, Pacific J. Math. 15, 1421-1427, 1965.

[T] J. Tits, *Une classe d'algebres de Lie en relation avec les algebres de Jordan*, Indag. Math. 24, 530-535, 1962.

[Z1] E. Zelmanov, *Jordan Algebras with finiteness conditions*, Algebra i Logika,1978.

[Z2] E. Zelmanov, *Absolute zero divisors and algebraic Jordan algebras*, SIberian Math. J. 23, 1982.

[Z3] E. Zelmanov, *On prime Jordan algebras II*, Siberian Math. J. 24, 1983.

[Z4] E. Zelmanov, *Characterization of the McCrimmon radical*, Sibirsk. Math. Zh. 25, 190-192, 1984.

[Z5] E. Zelmanov, *Lie algebras with a finite grading*, Math. USSR Sbornik 52(2),1985

[Z6] E. Zelmanov, *The solution of the Restricted Burnside problem for Groups of Odd Exponent*, Math. USSR Izv 36, 41-60, 1991.

[ZS] K. A. Zhevlakov and I.P. Shestakov, *On local finiteness in Shirshov's sense*, Algebra i Logika 12, N.1, 41-73, 1973.

[ZSSS] K.A. Zhevlakov, A.M.Slin'ko, I.P.Shestakov, A.I.Shirshov, *Rings that are nearly associative*, Academic Press, New York, 1982.

Editorial Information

To be published in the *Memoirs*, a paper must be correct, new, nontrivial, and significant. Further, it must be well written and of interest to a substantial number of mathematicians. Piecemeal results, such as an inconclusive step toward an unproved major theorem or a minor variation on a known result, are in general not acceptable for publication. Papers appearing in *Memoirs* are generally longer than those appearing in *Transactions*, which shares the same editorial committee.

As of November 30, 2000, the backlog for this journal was approximately 10 volumes. This estimate is the result of dividing the number of manuscripts for this journal in the Providence office that have not yet gone to the printer on the above date by the average number of monographs per volume over the previous twelve months, reduced by the number of volumes published in four months (the time necessary for preparing a volume for the printer). (There are 6 volumes per year, each containing at least 4 numbers.)

A Consent to Publish and Copyright Agreement is required before a paper will be published in the *Memoirs*. After a paper is accepted for publication, the Providence office will send a Consent to Publish and Copyright Agreement to all authors of the paper. By submitting a paper to the *Memoirs*, authors certify that the results have not been submitted to nor are they under consideration for publication by another journal, conference proceedings, or similar publication.

Information for Authors

Memoirs are printed from camera copy fully prepared by the author. This means that the finished book will look exactly like the copy submitted.

The paper must contain a *descriptive title* and an *abstract* that summarizes the article in language suitable for workers in the general field (algebra, analysis, etc.). The *descriptive title* should be short, but informative; useless or vague phrases such as "some remarks about" or "concerning" should be avoided. The *abstract* should be at least one complete sentence, and at most 300 words. Included with the footnotes to the paper should be the 2000 *Mathematics Subject Classification* representing the primary and secondary subjects of the article. The classifications are accessible from `www.ams.org/msc/`. The list of classifications is also available in print starting with the 1999 annual index of *Mathematical Reviews*. The Mathematics Subject Classification footnote may be followed by a list of *key words and phrases* describing the subject matter of the article and taken from it. Journal abbreviations used in bibliographies are listed in the latest *Mathematical Reviews* annual index. The series abbreviations are also accessible from `www.ams.org/publications/`. To help in preparing and verifying references, the AMS offers MR Lookup, a Reference Tool for Linking, at `www.ams.org/mrlookup/`. When the manuscript is submitted, authors should supply the editor with electronic addresses if available. These will be printed after the postal address at the end of the article.

Electronically prepared manuscripts. The AMS encourages electronically prepared manuscripts, with a strong preference for AMS-LaTeX. To this end, the Society has prepared AMS-LaTeX author packages for each AMS publication. Author packages include instructions for preparing electronic manuscripts, the *AMS Author Handbook*, samples, and a style file that generates the particular design specifications of that publication series. Though AMS-LaTeX is the highly preferred format of TeX, author packages are also available in AMS-TeX.

Authors may retrieve an author package from e-MATH starting from `www.ams.org/tex/` or via FTP to `ftp.ams.org` (login as `anonymous`, enter username as password, and type `cd pub/author-info`). The *AMS Author Handbook* and the *Instruction Manual* are available in PDF format following the author packages link from `www.ams.org/tex/`. The author package can be obtained free of charge by sending email to `pub@ams.org` (Internet) or from the Publication Division, American Mathematical Society, P.O. Box 6248, Providence, RI 02940-6248. When requesting an author package, please specify $\mathcal{A}\!\mathcal{M}\!\mathcal{S}$-LaTeX or $\mathcal{A}\!\mathcal{M}\!\mathcal{S}$-TeX, Macintosh or IBM (3.5) format, and the publication in which your paper will appear. Please be sure to include your complete mailing address.

Sending electronic files. After acceptance, the source file(s) should be sent to the Providence office (this includes any TeX source file, any graphics files, and the DVI or PostScript file).

Before sending the source file, be sure you have proofread your paper carefully. The files you send must be the EXACT files used to generate the proof copy that was accepted for publication. For all publications, authors are required to send a printed copy of their paper, which exactly matches the copy approved for publication, along with any graphics that will appear in the paper.

TeX files may be submitted by email, FTP, or on diskette. The DVI file(s) and PostScript files should be submitted only by FTP or on diskette unless they are encoded properly to submit through email. (DVI files are binary and PostScript files tend to be very large.)

Electronically prepared manuscripts can be sent via email to `pub-submit@ams.org` (Internet). The subject line of the message should include the publication code to identify it as a Memoir. TeX source files, DVI files, and PostScript files can be transferred over the Internet by FTP to the Internet node `e-math.ams.org` (130.44.1.100).

Electronic graphics. Comprehensive instructions on preparing graphics are available at `www.ams.org/jourhtml/graphics.html`. A few of the major requirements are given here.

Submit files for graphics as EPS (Encapsulated PostScript) files. This includes graphics originated via a graphics application as well as scanned photographs or other computer-generated images. If this is not possible, TIFF files are acceptable as long as they can be opened in Adobe Photoshop or Illustrator. No matter what method was used to produce the graphic, it is necessary to provide a paper copy to the AMS.

Authors using graphics packages for the creation of electronic art should also avoid the use of any lines thinner than 0.5 points in width. Many graphics packages allow the user to specify a "hairline" for a very thin line. Hairlines often look acceptable when proofed on a typical laser printer. However, when produced on a high-resolution laser imagesetter, hairlines become nearly invisible and will be lost entirely in the final printing process.

Screens should be set to values between 15% and 85%. Screens which fall outside of this range are too light or too dark to print correctly. Variations of screens within a graphic should be no less than 10%.

Inquiries. Any inquiries concerning a paper that has been accepted for publication should be sent directly to the Electronic Prepress Department, American Mathematical Society, P. O. Box 6248, Providence, RI 02940-6248.

Editors

This journal is designed particularly for long research papers, normally at least 80 pages in length, and groups of cognate papers in pure and applied mathematics. Papers intended for publication in the *Memoirs* should be addressed to one of the following editors. In principle the Memoirs welcomes electronic submissions, and some of the editors, those whose names appear below with an asterisk (*), have indicated that they prefer them. However, editors reserve the right to request hard copies after papers have been submitted electronically. Authors are advised to make preliminary email inquiries to editors about whether they are likely to be able to handle submissions in a particular electronic form.

Algebra to CHARLES CURTIS, Department of Mathematics, University of Oregon, Eugene, OR 97403-1222 email: `cwc@darkwing.uoregon.edu`

Algebraic geometry and commutative algebra to LAWRENCE EIN, Department of Mathematics, University of Illinois, 851 S. Morgan (M/C 249), Chicago, IL 60607-7045; email: `ein@uic.edu`

Algebraic topology and cohomology of groups to STEWART PRIDDY, Department of Mathematics, Northwestern University, 2033 Sheridan Road, Evanston, IL 60208-2730; email: `priddy@math.nwu.edu`

Combinatorics and Lie theory to SERGEY FOMIN, Department of Mathematics, University of Michigan, Ann Arbor, Michigan 48109-1109; email: `fomin@math.lsa.umich.edu`

Complex analysis and complex geometry to DUONG H. PHONG, Department of Mathematics, Columbia University, 2990 Broadway, New York, NY 10027-0029; email: `dp@math.columbia.edu`

*__Differential geometry and global analysis__ to LISA C. JEFFREY, Department of Mathematics, University of Toronto, 100 St. George St., Toronto, ON Canada M5S 3G3; email: `jeffrey@math.toronto.edu`

*__Dynamical systems and ergodic theory__ to ROBERT F. WILLIAMS, Department of Mathematics, University of Texas, Austin, Texas 78712-1082; email: `bob@math.utexas.edu`

Geometric topology, knot theory, hyperbolic geometry, and general topoogy to JOHN LUECKE, Department of Mathematics, University of Texas, Austin, TX 78712-1082; email: `luecke@math.utexas.edu`

Harmonic analysis, representation theory, and Lie theory to ROBERT J. STANTON, Department of Mathematics, The Ohio State University, 231 West 18th Avenue, Columbus, OH 43210-1174; email: `stanton@math.ohio-state.edu`

*__Logic__ to THEODORE SLAMAN, Department of Mathematics, University of California, Berkeley, CA 94720-3840; email: `slaman@math.berkeley.edu`

Number theory to MICHAEL J. LARSEN, Department of Mathematics, Indiana University, Bloomington, IN 47405; email: `larsen@math.indiana.edu`

Operator algebras and functional analysis to BRUCE E. BLACKADAR, Department of Mathematics, University of Nevada, Reno, NV 89557; email: `bruceb@math.unr.edu`

*__Ordinary differential equations, partial differential equations, and applied mathematics__ to PETER W. BATES, Department of Mathematics, Brigham Young University, 292 TMCB, Provo, UT 84602-1001; email: `peter@math.byu.edu`

*__Partial differential equations and applied mathematics__ to BARBARA LEE KEYFITZ, Department of Mathematics, University of Houston, 4800 Calhoun Road, Houston, TX 77204-3476; email: `keyfitz@uh.edu`

*__Probability and statistics__ to KRZYSZTOF BURDZY, Department of Mathematics, University of Washington, Box 354350, Seattle, Washington 98195-4350; email: `burdzy@math.washington.edu`

*__Real and harmonic analysis and geometric partial differential equations__ to WILLIAM BECKNER, Department of Mathematics, University of Texas, Austin, TX 78712-1082; email: `beckner@math.utexas.edu`

All other communications to the editors should be addressed to the Managing Editor, WILLIAM BECKNER, Department of Mathematics, University of Texas, Austin, TX 78712-1082; email: `beckner@math.utexas.edu`.

Selected Titles in This Series

(Continued from the front of this publication)

681 **David P. Blecher, Paul S. Muhly, and Vern I. Paulsen,** Categories of operator modules (Morita equivalence and projective modules, 2000

680 **Joachim Zacharias,** Continuous tensor products and Arveson's spectral C^*-algebras, 2000

679 **Y. A. Abramovich and A. K. Kitover,** Inverses of disjointness preserving operators, 2000

678 **Wilhelm Stannat,** The theory of generalized Dirichlet forms and its applications in analysis and stochastics, 1999

677 **Volodymyr V. Lyubashenko,** Squared Hopf algebras, 1999

676 **S. Strelitz,** Asymptotics for solutions of linear differential equations having turning points with applications, 1999

675 **Michael B. Marcus and Jay Rosen,** Renormalized self-intersection local times and Wick power chaos processes, 1999

674 **R. Lawther and D. M. Testerman,** A_1 subgroups of exceptional algebraic groups, 1999

673 **John Lott,** Diffeomorphisms and noncommutative analytic torsion, 1999

672 **Yael Karshon,** Periodic Hamiltonian flows on four dimensional manifolds, 1999

671 **Andrzej Rosłanowski and Saharon Shelah,** Norms on possibilities I: Forcing with trees and creatures, 1999

670 **Steve Jackson,** A computation of δ_5^1, 1999

669 **Seán Keel and James M^c^Kernan,** Rational curves on quasi-projective surfaces, 1999

668 **E. N. Dancer and P. Poláčik,** Realization of vector fields and dynamics of spatially homogeneous parabolic equations, 1999

667 **Ethan Akin,** Simplicial dynamical systems, 1999

666 **Mark Hovey and Neil P. Strickland,** Morava K-theories and localisation, 1999

665 **George Lawrence Ashline,** The defect relation of meromorphic maps on parabolic manifolds, 1999

664 **Xia Chen,** Limit theorems for functionals of ergodic Markov chains with general state space, 1999

663 **Ola Bratteli and Palle E. T. Jorgensen,** Iterated function systems and permutation representation of the Cuntz algebra, 1999

662 **B. H. Bowditch,** Treelike structures arising from continua and convergence groups, 1999

661 **J. P. C. Greenlees,** Rational S^1-equivariant stable homotopy theory, 1999

660 **Dale E. Alspach,** Tensor products and independent sums of $\mathcal{L}_p$-spaces, $1 < p < \infty$, 1999

659 **R. D. Nussbaum and S. M. Verduyn Lunel,** Generalizations of the Perron-Frobenius theorem for nonlinear maps, 1999

658 **Hasna Riahi,** Study of the critical points at infinity arising from the failure of the Palais-Smale condition for n-body type problems, 1999

657 **Richard F. Bass and Krzysztof Burdzy,** Cutting Brownian paths, 1999

656 **W. G. Bade, H. G. Dales, and Z. A. Lykova,** Algebraic and strong splittings of extensions of Banach algebras, 1999

655 **Yuval Z. Flicker,** Matching of orbital integrals on $GL(4)$ and $GSp(2)$, 1999

654 **Wancheng Sheng and Tong Zhang,** The Riemann problem for the transportation equations in gas dynamics, 1999

653 **L. C. Evans and W. Gangbo,** Differential equations methods for the Monge-Kantorovich mass transfer problem, 1999

For a complete list of titles in this series, visit the AMS Bookstore at **www.ams.org/bookstore/**.